EAI/Springer Innovations in Communication and Computing

Series Editor
Imrich Chlamtac, European Alliance for Innovation
Ghent, Belgium

The impact of information technologies is creating a new world yet not fully understood. The extent and speed of economic, life style and social changes already perceived in everyday life is hard to estimate without understanding the technological driving forces behind it. This series presents contributed volumes featuring the latest research and development in the various information engineering technologies that play a key role in this process. The range of topics, focusing primarily on communications and computing engineering include, but are not limited to, wireless networks; mobile communication; design and learning; gaming; interaction; e-health and pervasive healthcare; energy management; smart grids; internet of things; cognitive radio networks; computation; cloud computing; ubiquitous connectivity, and in mode general smart living, smart cities, Internet of Things and more. The series publishes a combination of expanded papers selected from hosted and sponsored European Alliance for Innovation (EAI) conferences that present cutting edge, global research as well as provide new perspectives on traditional related engineering fields. This content, complemented with open calls for contribution of book titles and individual chapters, together maintain Springer's and EAI's high standards of academic excellence. The audience for the books consists of researchers, industry professionals, advanced level students as well as practitioners in related fields of activity include information and communication specialists, security experts, economists, urban planners, doctors, and in general representatives in all those walks of life affected ad contributing to the information revolution.

Indexing: This series is indexed in Scopus, Ei Compendex, and zbMATH.

About EAI - EAI is a grassroots member organization initiated through cooperation between businesses, public, private and government organizations to address the global challenges of Europe's future competitiveness and link the European Research community with its counterparts around the globe. EAI reaches out to hundreds of thousands of individual subscribers on all continents and collaborates with an institutional member base including Fortune 500 companies, government organizations, and educational institutions, provide a free research and innovation platform. Through its open free membership model EAI promotes a new research and innovation culture based on collaboration, connectivity and recognition of excellence by community.

Lucia Knapcikova • Shakib Zohrehvandi
Editors

9th International Conference on Mobility, IoT and Smart Cities

EAI Mobility IoT 2022

Springer

EAI
RESEARCH MEETS INNOVATION

Editors
Lucia Knapcikova
Faculty of Manufacturing Technologies
Technical University of Košice
Prešov, Slovakia

Shakib Zohrehvandi
Kharazmi University
Tehran Province, Iran

ISSN 2522-8595 ISSN 2522-8609 (electronic)
EAI/Springer Innovations in Communication and Computing
ISBN 978-3-031-28224-9 ISBN 978-3-031-28225-6 (eBook)
https://doi.org/10.1007/978-3-031-28225-6

This Springer imprint is published by the registered company Springer Nature Switzerland AG
The registered company address is: Gewerbestrasse 11, 6330 Cham, Switzerland

Preface

The 9th International Conference on Mobility, IoT and Smart Cities: EAI Mobility IoT 2022 aims to bring together world-leading academics and practitioners from the fields of Industry 4.0 and IoT, urban mobility, Smart Cities, social mobility, and technological innovations. The unique combination of fields and disciplines focuses on Industry 4.0, the power of smarter information, and opportunities to create a bridge between science and practice. We are delighted to introduce the 9th edition of the 2022 European Alliance for Innovation (EAI) International Conference on Mobility, IoT and Smart Cities. The conference has brought researchers, developers, and practitioners worldwide. In light of the latest knowledge and findings from scientific projects, the authors present actual R&D trends in the given field. This issue defines the state of art in the area and explores related topics for future research.

The proceeding of EAI Mobility, IoT and Smart Cities 2022 consisted of 12 full papers. Aside from the high-quality technical paper presentations, the technical program also featured one keynote speech. We thank the Keynote Speaker, Dr. Ammar Odeh from Princess Sumaya University for Technology, Jordan, for his speech "Novel Algorithm for Secure Medical Images". We are also grateful to Conference EAI Manager Mikita Yelnitski for his support, all the authors who submitted their papers to the EAI Mobility, IoT and Smart Cities 2022 conference, and reviewers for their hard work.

Future cooperation will be as successful and stimulating as indicated by the contributions presented in this volume.

Prešov, Slovakia
Tehran, Iran

Lucia Knapcikova
Shakib Zohrehvandi

EAI Mobility IoT 2022: 9th EAI International Conference on Mobility, IoT and Smart Cities

December 15, 2022
Bratislava, Slovak Republic (Online)

Steering Committee

Assoc. Prof. Lucia Knapcikova
Technical University of Kosice, Slovak Republic

Shakib Zohrehvandi, Ph.D.,
Kharazmi University, Centre for European Studies, Iran

Organizing Committee

General Chairs

Assoc. Prof. Lucia Knapcikova
Technical University of Košice, Faculty of Manufacturing Technologies with a seat in Prešov, Slovakia

Shakib Zohrehvandi, Ph.D.,
Kharazmi University, Centre for European Studies, Iran

Technical Program Committee Chairs

Assoc. Prof. Lucia Knapcikova
Technical University of Košice, Faculty of Manufacturing Technologies with a seat in Prešov, Slovakia

Shakib Zohrehvandi, Ph.D.,
Kharazmi University, Centre for European Studies, Iran

Assoc. Prof. Roya Soltani
Khatam University, Tehran, Iran

Prof. Dragan Peraković
University of Zagreb, Croatia

Assist. Prof. Annamaria Behunova
Technical University of Košice, Faculty of Mining, Ecology, Process Control and Geotechnology, Slovakia

Web Chair

Assoc. Prof. Lucia Knapcikova
Technical University of Košice, Faculty of Manufacturing Technologies with a seat in Prešov, Slovakia

Publicity and Social Media Chairs

Assoc. Prof. Lucia Knapcikova
Technical University of Košice, Faculty of Manufacturing Technologies with a seat in Prešov, Slovakia

Shakib Zohrehvandi, Ph.D.,
Kharazmi University, Centre for European Studies, Iran

Workshop Chairs

Assoc. Prof. Lucia Knapcikova
Technical University of Košice, Faculty of Manufacturing Technologies with a seat in Prešov, Slovakia

Shakib Zohrehvandi, Ph.D.,
Kharazmi University, Centre for European Studies, Iran

Sponsorship & Exhibits Chair

Assoc. Prof. Lucia Knapcikova
Technical University of Košice, Faculty of Manufacturing Technologies with a seat in Prešov, Slovakia

Technical Program Committee Chairs

Shakib Zohrehvandi, Ph.D.
Kharazmi University, Centre for European Studies, Tehran, Iran

Ing. Lucia Knapčíková, Ph.D. Ing. Paed. IGIP
Technical University of Kosice, Faculty of Manufacturing Technologies, Slovak Republic

Contents

Part I
Social Mobility and Industry 4.0

Chapter 1
Utilization of Jakarta Kini (JAKI) Application to Support Smart Economy Implementation in DKI Jakarta

Rifqi Maajid and Rini Rachmawati

1 Introduction

The era of digital economic transformation nowadays is developing in line with the massive use of information and communication technology (ICT) in society. The use of ICT, which is also one of the main catalysts in this digital transformation, affects various aspects, one of which is urban planning. The Ministry of Industry is targeting to implement the "Making Indonesia 4.0" policy as a response to the current digital economy trend in Indonesia. The implementation of this policy is estimated to encourage real economic growth of 1–2% per year so that the growth of Indonesia's gross domestic product (GDP) will increase to 5% from 2018 to 2030 [1]. This is evidenced by the report of the Minister of National Development Planning that four digital sectors are projected to grow rapidly, namely, financial, technology, e-commerce, on-demand services, and the Internet of Things [2]. When viewed from the aspect of digitalization, Indonesia has become an extraordinary consumer of digital products, but it has not been used optimally in the context of economic digitization. This is an indication that Indonesia is facing a transition toward Industry 4.0, considering that the core of Industry 4.0 is a combination of production, information technology (ICT), and the Internet.

Based on research results [3], Indonesia is allegedly able to achieve growth of up to USD 150 billion or equivalent to 10% of GDP in 2025 if it succeeds in overcoming the problem of low internet penetration and utilization, as well as optimizing digital technology across sectors and expanding participation in the economic aspect. This is in line with the Digital Indonesia Roadmap 2021–2024,

R. Maajid (✉) · R. Rachmawati
Department of Development Geography, Faculty of Geography, Universitas Gadjah, Mada, Yogyakarta, Indonesia
e-mail: riqi.f.m@mail.ugm.ac.id; rinirachma@ugm.ac.id

L. Knapcikova, S. Zohrehvandi (eds.), *9th International Conference on Mobility, IoT and Smart Cities*, EAI/Springer Innovations in Communication and Computing,
https://doi.org/10.1007/978-3-031-28225-6_1

which is focused on four sectors, namely, digital infrastructure, digital government, digital economy, and digital society [4]. One of the keys to achieving these focuses is to prepare people to "go digital," where people are prepared to enter the digital space in various ways, one of which is through applications. The utilization of technology in this context is the main component of infrastructure development to realize smarter urban system services [5].

In general, each district/city in Indonesia has its application to support smart city development; in this case, DKI Jakarta, through the Jakarta Smart City Regional Public Service Agency (BLUD JSC), also launched a public service application under the DKI Provincial Government. Jakarta Smart City (JSC) implements the public service function in the smart city aspect through several indicators to measure the growth and development of the smart economy dimensions, including (1) utilization of e-commerce, (2) job creation, (3) one-stop payment integration, and (4) food security. A smart economy is characterized by the emergence of a digital economy system that is interconnected or integrated, thus causing efficient and effective relationships [6]. One of the goals of the smart economy is to ensure that the implementation of ICT in the transaction process (cashless) takes place in priority tourist areas and surrounding local governments [7]. Advances in ICT enable faster development and realization of several benefits such as new and digital products and services [8]. This is in line with one of the indicators to be achieved by JSC together with the DKI Provincial Government and other related parties in realizing one-stop payment integration to facilitate the transaction system carried out by the community. One form of implementing the smart economy is now integrated into the Jakarta Kini (JAKI) mobile application. The Jakarta Kini (JAKI) application functions as a platform for ICT penetration in supporting the success of digital government, digital society, and the digital economy in general, as well as one of the tangible forms of implementing the smart city dimension, including the smart economy dimension. This research aims to (1) identify the types of smart economy-based services that are integrated into the Jakarta Kini (JAKI) application and (2) analyze the use of the Jakarta Kini (JAKI) application in supporting the implementation of the smart economy in DKI Jakarta.

2 Method

This research was conducted using qualitative methods and presented in a qualitative descriptive. The data used in this research are quantitative and qualitative data sourced from secondary and primary data. Secondary data were obtained through the Jakarta Smart City (JSC) agency as the unit for developing the JAKI application and supporting data obtained from stakeholders and direct exploring through the JAKI application. Primary data were obtained through in-depth interviews and field observations. The informants were selected in in-depth interviews using non-probability sampling, namely, the snowball sampling technique, to deepen the analysis so that the data produced are more comprehensive. Data analysis techniques used in this study included qualitative descriptive, transcripts from in-depth interviews, data reduction, and triangulation. Triangulation is intended to test the credibility of data by checking data obtained from several sources. The data reduction results will be interpreted and analyzed following the research objectives.

3 Result and Discussion

3.1 Jakarta Kini (JAKI) Application

Jakarta Kini (JAKI) is an official public service and information center application for the DKI Jakarta Provincial Government developed by the Jakarta Smart City Regional Public Service Agency (BLUD). JAKI aims to meet the daily needs of Jakarta residents. JAKI combines features created by the government and the community (citizen-design services) and works as a digital service and official information from the DKI Jakarta Provincial Government (Fig. 1.1). JAKI is

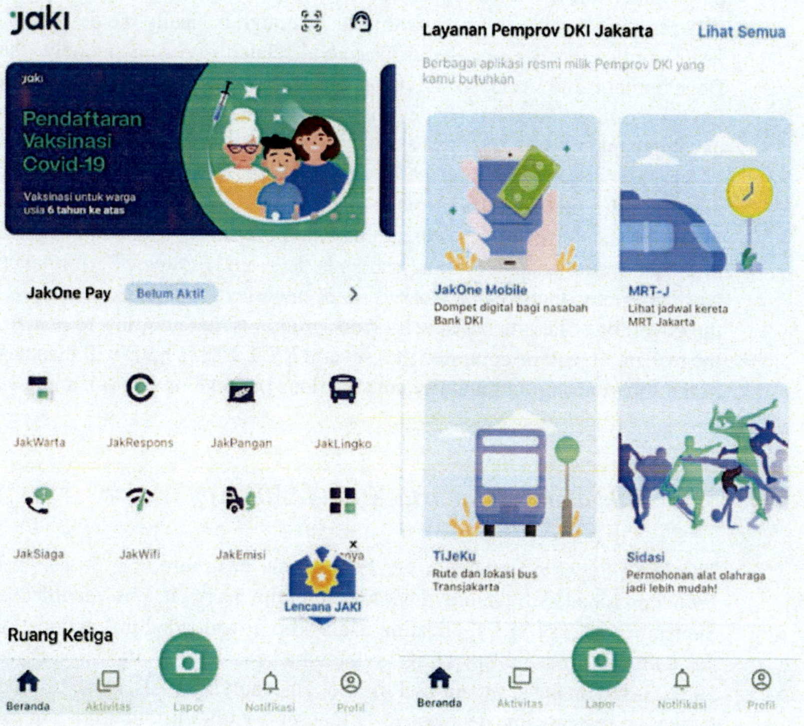

Fig. 1.1 The Jakarta Kini (JAKI) application dashboard

expected to become a super city app as well as a one-stop service aimed at Jakarta residents as part of a smart city to transform into a better city.

Initially, this application was launched for Android users only, but now the JAKI application can be used on Android and iOS OS devices. This is a manifestation of the accessibility principle upheld by the JAKI application. This principle means that JAKI must be able to reach all citizens from all walks of life in DKI Jakarta. This is also supported by the increasing number of smartphone users among Jakarta residents, so this application must be accessible from various operating system (OS) devices owned by Jakarta residents. In addition, JAKI also has a website (https://jaki.jakarta.go.id/en/) which users can freely access to find out what types of features or services are provided by JAKI.

Jakarta Smart City (JSC), as the developer of the JAKI application, structurally consists of several parts, which include the state civil apparatus (ASN), managers, experts, and intern staff. JAKI was developed by several divisions, with several divisions operating under the implementing unit, namely, the Data Analysis Division, Product Development and Analysis Division (PAP), Communications Division, Data Center and Network Division, Marketing Division, and Product and Service Division. The work system of each division in carrying out each participant is regulated under an activity/target to be achieved by the implementing unit. The development of the JAKI application cannot be separated from the collaboration between the relevant divisions within it. Although basically the development of the JAKI application is emphasized in several related divisions, namely, the Product Development and Analysis Division and the Product and Service Development Division, other divisions have contributed to the success of this application development.

The services provided at JAKI are measured based on their respective KPIs regarding the function and purpose of these services for use by users. In contrast, the types and forms of services are adjusted according to the applicable Regional Strategic Performance (KSD) adjustments. Regional Strategic Performance contains work programs that will be worked on in the relevant year. For example, by 2022, the governor of Jakarta has several work programs that he wants to realize, and then the realization will be compiled into several KSD's. This KSD will then be discussed in the Musrenbang (community consultations on development planning).

3.2 *Utilization of Jakarta Kini (JAKI) Application*

The JAKI Super App version 1.1.29 currently has more than 56 services and has been downloaded by more than four million users. It was recorded that as of September 2021, JAKI application users who downloaded this application through the iOS and Android operating systems had reached 4,430,388 users as of June 2022. Based on the calculation of the Google Analytics tools, it can be identified that there is an increase and decrease in "total users" with "current users" for each type of operating system (Fig. 1.2).

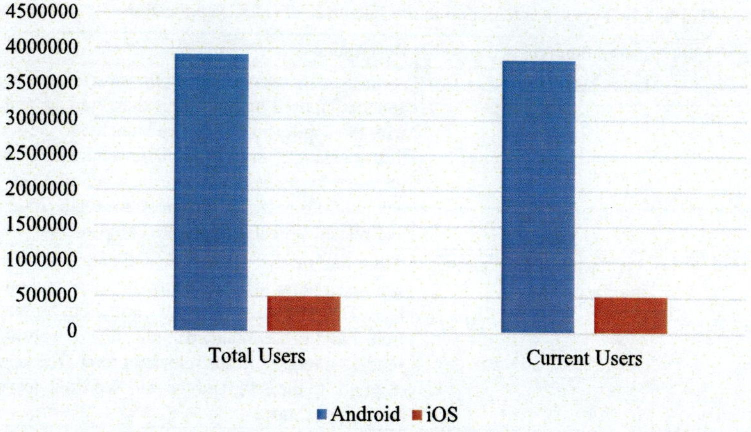

Fig. 1.2 A Comparison of total users with current users on the Jakarta Kini (JAKI) application

"Total users" are the accumulated number of users of the JAKI application as of 01/09/2019. Besides, "current users" is the total accumulation of old users plus new users detected by the Google Analytics system. It can be identified that among users of the Android operating system, it has decreased with the amount of difference between "total users" and "current users" is as many as 67,814 users. This means that as many as 67,814 people have installed the JAKI application, resulting in a reduction in the number of users. This is also influenced by the decreasing level of spread and impact generated by COVID-19 in 2022, considering that one of the most widely used services in the JAKI application is related to the handling of COVID-19, which is integrated into "Jakarta Response to COVID-19." It can be identified through the DKI Jakarta Provincial Government's policy to extend the Implementation of Community Activity Restrictions (PPKM) level 1 for the Jabodetabek area referring to the Instruction of the Minister of Home Affairs (Inmendagri) Number 26 of 2022 concerning the Enforcement of Community Activity Restrictions (PPKM) until July of 2018. 2022. It is different for users of the Android operating system. Users of the iOS system experience an increase in the number of new users with the difference between "total users" and "current users" of 12,907 users. Of course, many variables influence this increase in numbers, one of which is the shift in application users who replace the operating system on their respective gadgets. In this case, one of the things that can be considered.

In addition to calculating the number of users, the use of Google Analytics tools can also measure the value of the conversion rate. Conversion rate is a percentage of the number of website visitors who convert or take any action that benefits the business owner. In this context, the conversion rate can be interpreted as all activities carried out by users in the JAKI application that benefit Jakarta Smart City (JSC) as the business owner/party responsible for the development aspect of the application. The conversion rate, in this case, is divided into two classifications of Android and iOS operating systems, where the percentage of conversion value in the Android

Table 1.1 Types of integration in the Jakarta Kini (JAKI) application

No.	Integration type	Description
1.	Launcher	A mechanism used to launch the selected application in the application used by the user is then used for authentication with the application [10]. The launcher in the JAKI application is generally used to promote certain services or services provided by partners who are Jakarta Smart City (JSC) partners, both from the government, local government agencies, and related agencies to private companies.
2.	Application Programming Interface (API)	A mechanism that allows two software tools to communicate with each other using a set of definitions and protocols [11]. The API work system is carried out by sending access from JAKI to the intended application or website and then by sending a request to the relevant server. The server then responds to the API, and the API sends the response to the JAKI application.
3.	Banner Ads	A mechanism of "branding" or promotion is carried out by collaborators affiliated with JSC or the JAKI application, for example, local government agencies, and related agencies.
4.	WebView	A mechanism that allows developers to display the content of the destination website as part of the activity layout in the application [12]. The WebView in the JAKI application is used to direct the services in the application to the destination websites.

operating system is greater than the iOS operating system with a ratio of 88%: 12% (148,251,641: 20,325,629). This is influenced by the number of JAKI users using the Android operating system, which is much larger than the iOS operating system users. Hence, the conversion value is also much greater. The JAKI application does not yet have features/services in the form of "native" or specifically built for the JAKI application and can only be used in that application. This is because all JAKI is directed to be made in the form of integration with other applications or collaboration. These integration rules are regulated in [9], discussing procedures and procedures for integrating applications into the JAKI platform, which includes integration of JAKI with collaborators, integration of JAKI with internal parties of the Provincial Government, and integration of applications/platforms on JAKI. Based on the technical SOPs, the integration of JAKI can be divided into four types (Table 1.1).

Several types of smart economy integrated into the JAKI application are a tangible form of these goals (Table 1.2).

The types of integration of each of these services are in the form of Launcher, WebView, and Application Programming Interface (API). Sequentially, the development of the smart economy dimension, which is divided into four objectives, is then shortened into several terms, namely, "Collaboration to Build Jakarta," "Business with JakPreneur" as well as information on job vacancies via JakNaker, JakOne Pay, and dan JakPangan.

Table 1.2 Integration of smart economy-based services in the Jakarta Kini (JAKI) application

No.	Collaboration	Service	Integration Type
1.	Collaboration to Build Jakarta	1. JAKI X Bukalapak 2. JAKI X Duithape 3. JAKI X Gojek 4. JAKI X Grab 5. JAKI X Shopee 6. JAKI X Tokopedia	Launcher
2.	Business with JakPreneur	JakPreneur	WebView
3.	Job Vacancy in Jakarta	JakNaker	Banner Ads
4.	DKI Jakarta Provincial Government Services	JakOne Pay	1. Launcher 2. Application Programming Interface (API)
5.	JAKI #MakeItEasy	JakPangan	Application Programming Interface (API)

3.3 Implementation of Smart Economy in Jakarta Kini (JAKI) Application

Utilization of E-commerce

Utilization of e-commerce in this context is one part of the smart economy development plan that is echoed by Jakarta Smart City (JSC) as the DKI Jakarta BLUD, especially to increase the development of small and medium enterprises (SMEs) and MSMEs as well as increase competitiveness and expand the market for these businesses. This is also supported by the strengthening of MSMEs and innovations in products and marketing due to increased ICT utilization in the community [13]. JAKI as a super app, in this case, covers the existing demand related to the need for e-commerce development. JAKI integrates various DKI Jakarta Provincial Government websites that residents need and are already available and is projected as an integrator or entry gate for residents to access all DKI Jakarta Provincial Government online services through the "Collaboration to Build Jakarta" integration model.

"Collaboration to Build Jakarta" is one of the integration services in the JAKI application. This service is integrated using the launcher system. As for now, there are services provided with their respective purposes and functions. This includes several large e-commerce or marketplaces in Indonesia, including Tokopedia, Bukalapak, Shopee, Gojek, and Grab, as one the collaborators in this integration. JAKI's collaboration with major e-commerce is a form of collaboration that is in line with the "Spirit of Collaboration," which is carried as one of JAKI's slogans, so it can be said that applications integrated within JAKI can be categorized as part of

JAKI. However, until now, JAKI has not provided the type of integration that allows the sale and purchase scheme to occur with the e-commerce/marketplace cooperation.

One of the goals of attracting large e-commerce companies is to attract a large market of e-commerce users to use the JAKI application. This is in line with one of the visions of the JAKI super-app to become an "all-in-one" integrative application by reaching applications that have a large user intensity, especially in the DKI Jakarta area. Apart from that, other applications are also included in the "Collaboration to Build Jakarta," including NodeFlux, DuitHape, Safe Jakarta, SekolahMu, Molecool, and Google Maps (Table 1.3). This is one form of manifestation of the jargon "Spirit of Collaboration," which is always promoted and disseminated by Jakarta Smart City (JSC) as the application developer.

The types of integration together with large e-commerce have an objective goal of digitalization and digital economic growth in the community, especially for business actors in DKI Jakarta, ranging from SMEs to MSMEs. The forms of cooperation carried out tend to be diverse, ranging from training and mentoring to providing easy access to pay for entrance tickets to tourist destinations. This, of course, cannot be separated from the cooperation of third parties with related agencies and local government agencies. However, these activities cannot be categorized as "permanent activities" that are carried out continuously. So far, the JAKI application only offers a launcher integration type for users to directly access the landing page of the destination e-commerce integrated into this service. The developer, in this case, hopes to innovate so that full integration can be carried out so that the efficient use of applications according to the "super app" function can be implemented in the future or, in other words, can be perfectly integrated.

Job Creation

The increasing job opportunities by expanding job opportunities aimed at all people of DKI Jakarta, especially those who have the status of "workforce," is one of the goals to be achieved by the DKI Jakarta Provincial Government. Job creation then becomes one of the missions that the DKI Jakarta Regional Government wants to realize as a form of implementation of the smart economy roadmap. This expansion of employment opportunities is not solely to meet market demand for workforce needs. Still, more than that, it is focused on producing quality workers, having qualified skills and expertise and being able to encourage existing entrepreneurial sectors.

Based on [14], the purpose of the Integrated Entrepreneurship Development program is to create new entrepreneurs in the DKI Jakarta Province. The direction of job creation is to increase job openings that are intended not only for residents of DKI Jakarta but also with a very high level of urbanization in the capital. It is also hoped that it can also reach out to immigrants so that they can develop and have competitive competitiveness in the field of entrepreneurship, as well as adapt to all changes in economic conditions. One tangible form of implementing this mission is found in the JakPreneur and JakNaker services in the JAKI application.

Table 1.3 "Collaboration to Build Jakarta" services

No.	Integration type	Description
1.	JAKI X Bukalapak 	Encouraging a digital-based populist economy and supporting economic transparency by opening economic opportunities. This collaboration includes: 1. Digitizing fostered MSMEs to become an entrepreneur 2. Training of assisted MSMEs 3. Development of co-working space
2.	JAKI X Shopee 	Digital training for fostered MSMEs in DKI Jakarta. This collaboration includes: 1. MSME training 2. Creation of "Archipelago Creative Channel" 3. Purchase of entrance tickets to tourist destinations managed by the DKI Jakarta Tourism and Culture Office.
3.	JAKI X Gojek 	Encouraging the realization of sustainable mobility in DKI Jakarta. This collaboration includes: 1. Collaboration for the development of fostered MSMEs 2. Development of MSME digital literacy 3. POI (point of interest) information which is used for the driver's deposition area as a drop off and pick up point
4.	JAKI X Grab 	Supporting the integration of transportation and the digital economy in DKI Jakarta in various sectors. This collaboration includes: 1. Training of assisted MSMEs to become Grab Food partners 2. Provision of Grab Wheels at tourist sites 3. Provision of a driver's shelter for resting
5.	JAKI X Tokopedia 	Encouraging the growth of a sustainable digital economy ecosystem and ease of service for the people of DKI Jakarta. This collaboration includes: 1. Access to make PBB payments 2. Access to make eSamsat payments eSamsat 3. Training for MSMEs
6.	JAKI X Duithape 	Provide distribution of assistance such as zakat, KJP Plus, and KJMU to the community more quickly, precisely, and effectively. DKI X DuitHape Provincial Government cooperation includes: 1. KJP cashless distribution activities 2. Distribution of zakat 3. JakLingko digital payment
7.	JAKI X Jakarta Aman 	Creating a safe and comfortable city of Jakarta for the community with the panic button feature and a safe community with neighbors.

(continued)

Table 1.3 (continued)

No.	Integration type	Description
8.	JAKI X Nodeflux 	Encouraging the creation of services that utilize artificial intelligence technology. This collaboration includes: 1. Vehicle counting 2. Public mobility monitoring 3. Social distance monitoring 4. Face mask monitoring
9.	JAKI X Sekolahmu 	Provides a digital platform for schools to access educational content during the COVID-19 pandemic for free.
10.	JAKI X Google Maps 	Provides a shortcut to the Google Maps application to find alternative routes in the user's journey.
11.	JAKI X Molecool 	Provides free Wi-Fi connection information in DKI Jakarta that is integrated with JakWifi.

Jakpreneur is a platform for the creation, facilitation, and collaboration of MSME development that aims to develop skills and business independence creatively and collaboratively (Fig. 1.3). Jakpreneur is integrated with the JAKI Application in a WebView. The service called "Business with Jakpreneur" in the JAKI application allows users to access a display that is integrated with the Jakpreneur official website so that all data entered into the JAKI application will be automatically inputted and recorded by the JakPreneur database.

As of October 2022, JakPreneur services have been followed by 331,102 entrepreneurs. This number has exceeded the 2022 target, which is targeted for 200 thousand registered and fostered MSMEs following the political promise of the governor of DKI Jakarta through the Integrated Entrepreneurship Program, which was declared on December 14, 2017. The JakPreneur service, which is integrated into the JAKI application, offers several facilities that are divided into several stages, including (1) training, (2) assistance, (3) licensing, (4) marketing, and (5) financial reporting and capital facilities [15]. Jakpreneur is currently still in the form of a WebView. In the future, it is hoped that there will be an initiation to integrate users

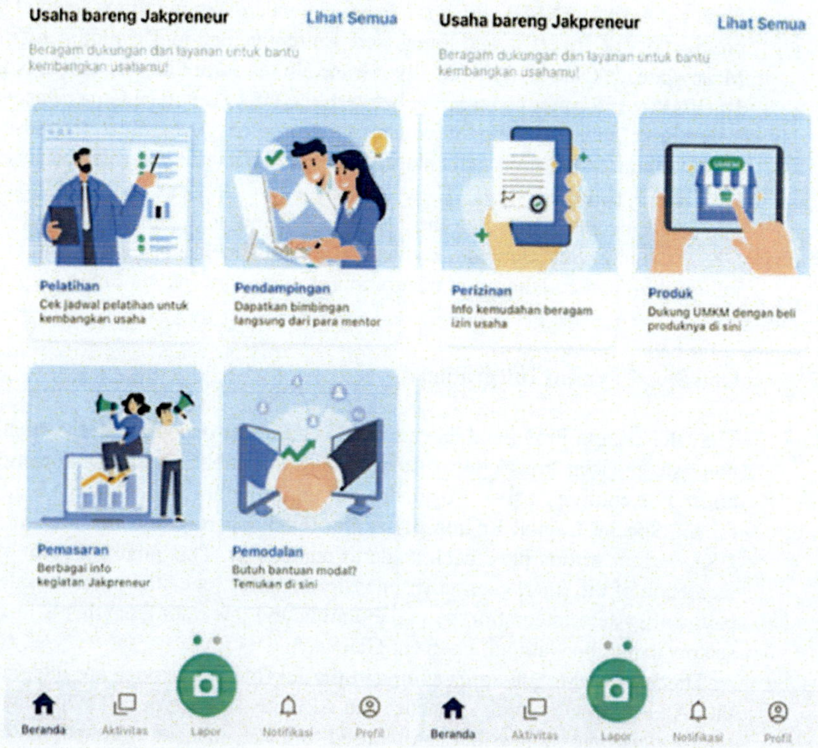

Fig. 1.3 Display of JakPreneur service integration in the JAKI application

with single logins. Single login allows users to log in once for two integrated applications.

Another feature that is in line with the development of smart economy indicators is job creation, which is JakNaker. The DKI Jakarta Provincial Government Cooperation Service, together with the DKI Jakarta Provincial Manpower, Transmigration and Energy Office, and Top Karir Indonesia, is a career portal intended for Jakarta residents to develop their careers. In general, the objective of providing JakNaker services has several functions, namely, (1) preparing superior human resources (HR) and ready to compete in the industrial world, (2) advancing the regional economy through MSMEs in DKI Jakarta, and (3) providing employment opportunities to reduce unemployment in DKI Jakarta.

Unlike the JakPreneur service which already has its own integration segment, JakNaker is currently only integrated with JAKI through Banner Ads in the JAKI application which is relatively new and still being developed. This is also influenced because JakPreneur itself has become its own application (native) as one of the answers to the DKI Jakarta Provincial Government in responding to the problem of unemployment and the high poverty rate in DKI Jakarta, as well as a Regional

Strategic Activity (KSD) which is a form of standardization improvement: OK OCE program work in 2018 and Integrated Entrepreneurship Development in 2019. Meanwhile, JakNaker was officially released in the form of a website on November 18, 2021, as a form of cooperation between the DKI Provincial Government and PT Top Karir Indonesia, one of which is a form of government response to the phenomenon of the high percentage of people affected by layoffs because of the COVID-19 pandemic [16]. Therefore, these two services, both JakPreneur and JakNaker, are integrated into the JAKI application as a form of support from the DKI Provincial Government in implementing the dimensions of the smart economy: job creation and collaboration integration promoted by Jakarta Smart City (JSC).

One-Stop Payment Integration

The DKI Jakarta Provincial Government is currently encouraging the creation of a noncash financial transaction system among the public (less cash society) that is holistic or comprehensive. Together with the Regional Development Bank (BPD) for the Special Capital Region of Jakarta (DKI Jakarta) or better known as Bank DKI, various efforts have been made to realize this. This goal is a form of implementation of the smart economy dimension, namely, one-stop payment integration/ payment system integration by collaborating and integrating payments from multi-sectors using the National Payment Gateway.

The JakOne Mobile application and the JakOne Pay service are integrated with the JAKI application with Launcher and API integration types. JakOne Pay is Bank DKI's electronic money that can be used to make financial transactions (Fig. 1.4). One of the advantages of using JakOne Pay is that the public or users of the JAKI application do not have to be customers of Bank DKI to be able to use the service. The integration of a JAKI account with JakOne Mobile to use the JakOne Pay service is carried out by entering personal data in the form of a cellphone number, PIN, and copying the OTP code sent via SMS inbox. The provision of this service is one of the efforts of Bank DKI and the DKI Provincial Government to support the normalization of cashless transaction services, especially concerning the MSME sector, which has been digitized. JakOne Pay is integrated with the Quick Response Code Indonesian Standard (QRIS) feature, a combination of various types of QR codes that are used to make digital transactions with various payment method systems or PJSP [17]. It was recorded that transactions using QRIS until the period of May 2022 through JakOne Mobile grew 561% (YoY) or quantitatively increased by Rp. 13 billion from the period of May 2021 of Rp. 2 billion [18].

JakOne Pay in the JAKI application provides three (3) functions, namely, the pay function, the top-up function, and the history function. The pay function is used to make transactions through QRIS, the top-up function is used to increase the balance to be filled to make transactions, and the history function is used to check the history of transactions that have been made. All these functions will be API integrated with the JakOne Mobile application so that data recording will be automatically connected and synchronized. The ease of making payments for goods is one of the

Fig. 1.4 Display of JakOne Mobile and JakOne Pay services integration in the JAKI Application

targets to be achieved in the development of smart economy indicators. The DKI Jakarta Provincial Government has so far called for adding other collaborators in the integration of digital wallets/digital payments by 2022.

Food Security

The aspect of food security is one of the important aspects that is one of the main priorities of the national development program because food has a strategic value that will also affect the sustainability of aspects of social, economic, and even political security. This is a special concern for various government apparatus in Indonesia, including the DKI Jakarta Provincial Government. Food availability is the condition of the availability of food produced in the country and national food reserves as well as imports if the two main sources cannot meet the needs [19]. The DKI Provincial Government then adopted this food security goal by focusing of them on disseminating information and providing food commodity price stock data in real time and easy to access [20]. It is hoped that the stock price and type of food data will be updated automatically following market share and prevailing market prices to facilitate public access to finding information related to the availability of food or markets, especially in the DKI Jakarta area.

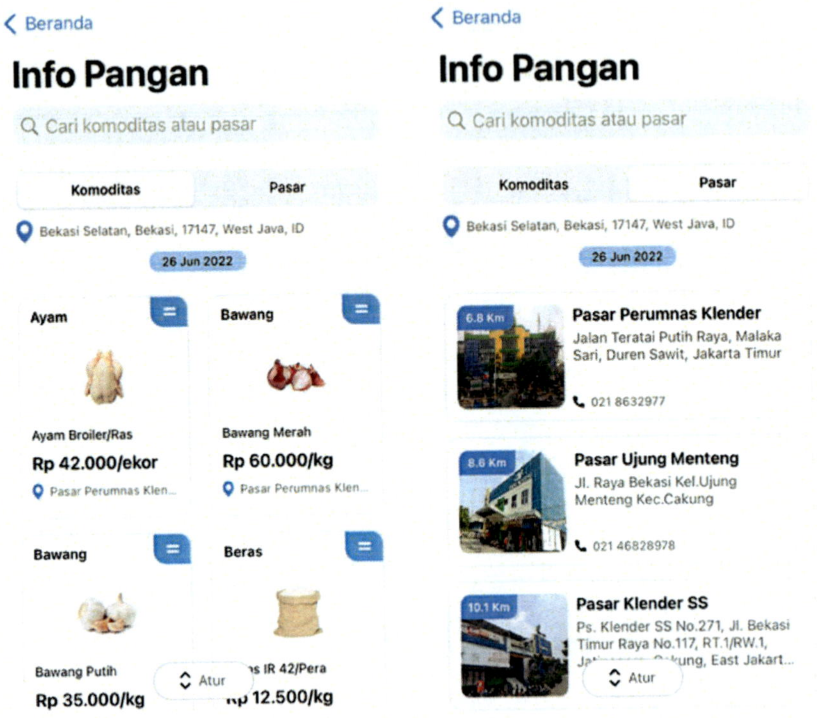

Fig. 1.5 Display of JakPangan service integration in the JAKI application

The DKI Provincial Government, through the BLUD Jakarta Smart City (JSC), launched the JakPangan service as a response to the focus on developing food security in Jakarta. JakPangan is an information service for food commodities in markets in Jakarta (Fig. 1.5). This service is integrated into the JAKI application with an integrated system in the form of an Application Programming Interface (API) whose data are sourced from the Jakarta Food Info (IPJ) website. The API integration allows users to get the latest updates from the Jakarta Food Info website, which is directly connected to the JAKI application. This integration was carried out in line with the development of the Strategic Food Price Information Center (PIHPS). The program is a form of food security strengthening program supported by the DKI Jakarta Regional Inflation Control Team (TPID), which includes members of the DKI Jakarta Provincial Government, PD Pasar Jaya, and Bank Indonesia.

In the future, JakPangan development is expected to not only be able to provide information related to access to food but also to enable users to make transactions directly and can place orders directly as a form of digital transformation facilitation. So far, Info Pangan Jakarta (IPJ), as an integrated source of data from JakPangan, has provided 37 commodity data and 48 market location data that JAKI users can access. As for the development of JakPangan, the community is indirectly expected to participate in monitoring existing food prices and is able to intervene in the event of a significant increase in food prices.

4 Conclusion

Based on the results, it can be concluded that JAKI is oriented toward three main values: integration, community orientation, and one-stop service. Smart economy-based services integrated into JAKI include five services: "Collaboration to Build Jakarta," JakPreneur, JakNaker, JakOne Pay, and JakPangan. The use of smart economy-based services in the JAKI application is integrated into four schemes: Launcher, Banner Ads, WebView, and API. Each available service is a form of collaboration between JAKI and various relevant stakeholders according to their respective functions and objectives. The service that is experiencing the fastest development in this regard is JakPreneur, which focuses on developing the entrepreneurial sector among the people of DKI Jakarta as well as a manifestation of one of the goals of developing a smart economy, namely, expanding employment opportunities by increasing skills and expertise according to market needs.

Acknowledgments This research was a part of Recognition of Final Student Project Batch I (Rekognisi Tugas Akhir/RTA Batch I) 2022 from Universitas Gadjah Mada, Indonesia led by Dr. Rini Rachmawati, S.Si., M.T. Therefore, the authors thank Universitas Gadjah Mada, which has supported this research activities and its funding.

References

1. Adiningsih, S. (2019). *Transformasi ekonomi berbasis digital di Indonesia: lahirnya tren baru teknologi, bisnis, ekonomi, dan kebijakan di Indonesia*. PT Gramedia Pustaka Utama.
2. Setyowati. (2019). *Empat Sektor Digital Diproyeksi Tumbuh Pesat hingga 2020*. https://katadata.co.id/pingitaria/digital/5e9a555252bd7/empat-sektor-digital-diproyeksi-tumbuh-pesat-hingga-2020
3. McKinsey. (2016). *Unlocking Indonesia's digital opportunity*. https://www.mckinsey.com/featured-insights/asia-pacific/unlocking-indonesias-digital-opportunity
4. Kominfo. (2021). *Menkominfo Paparkan Roadmap Digital Indonesia 2021–2024 di 4 Sektor Strategis*. https://diskominfo.natunakab.go.id/menkominfo-paparkan-roadmap-digital-indonesia-2021-2024-di-4-sektor-strategis
5. Rachmawati, R. (2018). Pengembangan Smart Village Untuk Penguatan Smart City Dan Smart Regency. *Jurnal Sistem Cerdas, 1*(2), 12–19.
6. Hoetoro, A., & Satria, D. (2020). *Smart Economy: Kewirausahaan UMKM 4.0*. UB Press.
7. Kominfo. (2019). *Perkembangan Ekonomi Digital di Indonesia Strategi dan Sektor Potensial*. (Badan Penelitian dan Pengembangan SDM Puslitbang Aptika dan IKP). https://balitbangsdm.kominfo.go.id/?mod=publikasi&cid=3&filter=format-hasil&page=1
8. Youssef, A., & Hajek, P. (2021). The role of smart economy in developing smart cities. *2021 International Symposium on Computer Science and Intelligent Controls (ISCSIC)*, 2021, pp. 276–279, https://doi.org/10.1109/ISCSIC54682.2021.00057.

9. Diskominfotik. (2021). Keputusan Kepala Dinas Komunikasi, Informatika, dan Statistik Provinsi Daerah Khusus Khusus Ibukota Jakarta Nomor 4 Tahun 2021 Tentang Pedoman Pengintegrasian Aplikasi Pada Platform Jakarta Kini (JAKI).
10. International Business Machines (IBM). (2019). *Launch application definitions*. https://www.ibm.com/docs/SSDKXQ_6.3.0/com.ibm.itm.doc/itm_tep251.htm
11. International Business Machines (IBM). (2022). *What is an Application Programming Interface (API)*. https://www.ibm.com/cloud/learn/api
12. Developers. (2022). *WebView*. developer.android.com
13. Rachmawati, R., Mei, E. T. W., Nurani, I. W., Ghiffari, R. A., Rohmah, A. A., & Sejati, M. A. (2021). Innovation in coping with the COVID-19 pandemic: The best practices from five smart cities in Indonesia. *Sustainability, 13*(21), 12072. https://doi.org/10.3390/su132112072
14. DKI Jakarta. (2020). Peraturan Gubernur DKI Jakarta No. 2 Tahun 2020 tentang Penyelenggaraan Pengembangan Kewirausahaan Terpadu (PKT).
15. Aditya, G., & S. Teresa, (2021). Jakpreneur: Platform Kolaborasi untuk Majukan UMKM Jakarta. https://smartcity.jakarta.go.id/id/blog/jakpreneur-platform-kolaborasi-untuk-majukan-umkm-jakarta/
16. TopKarir. (2021). *Peluncuran Website Karir Kolaboratif JakNaker.id untuk Berdayakan Talenta Muda Ibukota*. https://smk.topkarir.com/2/article/detail/peluncuran-website-karir-kolaboratif-jaknakerid-untuk-berdayakan-talenta-muda-ibukota
17. Regulation of Members of the Board of Governors Number 21/ 18 /PADG/2019 concerning Implementation of the Quick Response Code National Standard for Payments. (2019). https://www.bi.go.id/id/publikasi/peraturan/Pages/padg_211819.aspx
18. Arini, S. C. (2022). *Inilah JakOne Pay, Aplikasi Milik Bank DKI di Jakarta Formula E*, https://finance.detik.com/berita-ekonomi-bisnis/d-6111766/inilah-jakone-pay-aplikasi-milik-bank-dki-di-jakarta-formula-e-2022
19. Undang-. (2012). Undang Nomor 18 Tahun 2012 tentang Pangan.
20. DKI Jakarta. (2017). Peraturan Daerah Nomor 1 tahun 2018 tentang RPJMD Provinsi DKI Jakarta Tahun 2017–2022.

Chapter 2
Collaboration of an Established Company with Startups: A Funnel-Shaped Framework

Zahra Honarmand Shahzileh, Mohammad Aghdasi, and Amir Albadvi

1 Introduction

Access to cheaper and better information and communication technology has grown with digitization. As a result of digitization, established businesses are threatened by startups which emerge in their industry area forced them to identify innovation opportunities and launch new products faster in order to maintain and expand their competitive advantage [7, 74–77, 83]. Startup is a global topic. Startups are newborn entities that are striving for their survival and sustainability. These entities are mainly built on top of original ideas that can turn into successful stable companies in right conditions. There is no single definition of what constitutes a startup in the literature [57], and each author has their own interpretation of the concept [70]. Blank [5] defined startup as a temporary organization that seeks an expansible and replicable business model. McClure [51] defines startup as a company that is perplexed about three issues: what is it producing? Who are its customers? How can it make money? As soon as the startup figures all three out, it grows out of being a startup and turns into a real business. A startup may be regarded as the early stage of creating a venture or a new organization [68]. Avnimelech and Teubal [2] defined startups as young and high-tech companies that their main activity is to work with a new venture idea till the initial sale stage arrives which usually takes 1–5 years. Startups are organizations that produce new products under highly unreliable conditions [72]. Ries [65]

Z. H. Shahzileh (✉)
Tarbiat Modares University, Tehran, Iran

Lappeenranta-Lahti University of Technology, Lappeenranta, Finland
e-mail: zahra.honarmand@modares.ac.ir; Zahra.honarmand.shahzileh@lut.fi

M. Aghdasi · A. Albadvi
Tarbiat Modares University, Tehran, Iran

L. Knapcikova, S. Zohrehvandi (eds.), *9th International Conference on Mobility, IoT and Smart Cities*, EAI/Springer Innovations in Communication and Computing,
https://doi.org/10.1007/978-3-031-28225-6_2

defined startups as a human institution designed to create an innovative product/ service under substantial uncertainty. Startups are the "new instruments of production" [3]. Graham [24] defined startup as a company focused on "growing" that doesn't need to be newborn or working on a technology; rather, it relies on the investment budget and requires an exit strategy. Startups face many uncertainties because they do not have a complete business model, and most startup founders consider their work as a temporary project and treat it as such. Rather than insisting on business plan, original startups are a fan of replicating and releasing a product based on previous feedbacks [82]. Kopera et al. [38] define startups as a dynamism of form and merit that results from operating in a highly volatile and unpredictable environment. Corvello et al. defined startup as an "entrepreneurial venture" that is looking for financial backing to get off the ground; startups are distinguished by a scarcity of both human and financial resources, as well as experience [12]. Refinement of startup ecosystems has led to their increasing growth in the global scale [21]. It is important for most established companies to know how they can acquire innovations made in startups through a mutually beneficial collaboration. Several research works have been carried out about what criteria should the collaboration between a big company and startups be based upon. Kohler introduced a framework for designing corporate accelerators [37]. De Latour et al. [15] investigated the collaboration between a company and startups engaged in very deep technologies. Gutmann [26] proposed a 3*3 framework through which he intended to see which modes of corporate venturing and which dimensions of them can be seen in the literature and how they could be extensively categorized. Collaboration between large companies and startups brings various challenges for each of them since they have significant and almost opposite dissimilarities, yet these dissimilarities can be complementary. Five types of challenges are identified [52]: strategic and business model, technology, organization, deal arrangement, and deal management. Wang [82] extracted the main challenges facing startups: budget, team making, business model, excessive capacity/too much work to do, income, minimum viable product (MVP), staying focused and orderly, suitability of product for the market, critical mass, scaling, adjusting the solution to the problem, coordinating the leader with the team, contributing, and legal regulations [82].

Most corporations struggle to find a tailored collaboration mechanism. Furthermore, academic research on corporate–startup collaboration is fragmented, with competing frameworks and typologies. This paper is one of the first studies to explicitly address the phenomenon of collaboration between established corporations and startups from the perspective of the corporation. Our framework is a visual representation of how to design the process of collaboration with startups. The stakeholders involved in corporate–startup collaboration initiatives will find our framework useful, stakeholders such as corporations, startups, independent accelerators, incubators, governments, scholars, and practitioners. We will enrich the literature on corporate startup engagement by organizing and integrating prior literature and harmonizing competing approaches. We will introduce a coherent and reconciled funnel-shaped framework which scholars can use as a ground for producing hypothesis and generalizable findings.

This chapter is organized as follows. In next section, the methodology of the study is discussed, and in third section, our proposed framework of collaboration, its dimensions, and constructs are presented. Discussion and research restrictions are presented in the fourth section, further research in the fifth section, and finally, conclusion in the sixth section.

2 Research Methods

Reviewing the literature regarding the collaboration of a large company with startups helps to gain a deep understanding and vision about this topic. Therefore, the literature has been systematically reviewed in this study by using www.scopus.com database. Then, the findings of this literature review are categorized, and the framework is extracted. The search for corresponding sources was conducted in English. Three key phrases were searched, including startup and company collaboration, startup and company partnership and startup, and company engagement. All the results published before August 2022 were subject to our analysis. After searching, first the abstracts were read, and then if deemed appropriate, the whole paper would have been studied. A total of 298 sources were found while 93 abstracts were reviewed, and 42 papers were completely studied (Table 2.1).

3 Funnel-Shaped Framework

The funnel-shaped framework presented in this paper has two parts: "dimensions" and "constructs." Each part will be explained in details.

3.1 Dimensions of Collaboration Framework

We have considered three perspectives to study the cooperation between an established company and startups; accordingly, our framework has three dimensions: (1) established company, (2) startup, and (3) the collaboration process. The first dimension is established company. Here, the established company is a company with a well-defined business model that can work in any industry. The second dimension is startups. Based on Weiblen and Chesbrough's article [83], "it's the startups, and not the big companies, that come with big new things to create an unconditional market space and change the whole industry." Since startups are small, they suffer from shortage of tangible and intangible resources, but large firms have more operational resources and significant financial resources. The third dimension is the collaboration. Collaboration between two companies can be regarded as an alliance that can have various levels. Partnership agreement can be

Table 2.1 List of the sources, studied completely

No	Author	Title	Year
1	Varesco Kager N., Sparr J.L., Grote G.	Looking for guidance? Five principles for leveraging tensions in corporate–startup collaboration	2022
2	Gutmann, T. Lang, C.	Unlocking the magic of corporate-startup collaboration: How to make it work	2022
3	Hübel, C., Weissbrod, I., Schaltegger, S.	Strategic alliances for corporate sustainability innovation: The 'how' and 'when' of learning processes	2022
4	Corvello V., Steiber A., Alänge S.	Antecedents, processes and outcomes of collaboration between corporates and start-ups	2021
5	Cordes, A., Guderian, C.C., Riar, F.J.	Exploring the practice of evaluation in corporate venturing	2021
6	Urbaniec, M, and Agnieszka Ż.	Business model innovation in corporate entrepreneurship: Exploratory insights from corporate accelerators	2021
7	Kurpjuweit S., Wagner S.M., Choi T. Y.	Selecting startups as suppliers: A typology of supplier selection archetypes	2021
8	Reisdorfer-Leite, B. Michele Marcos de, O. Marcelo Rudek, A. Luis, S. and Osiris, C. J.	Startup definition proposal using product lifecycle management	2020
9	Steiber A., Alänge S.	Corporate-startup co-creation for increased innovation and societal change	2020
10	Steiber A., Alänge S.	Corporate-startup collaboration: Effects on large firms' business transformation	2020
11	Allmendingert, M. and Berger, E.	Selecting corporate firms for collaborative innovation: Entrepreneurial decision making in asymmetric partnerships.	2020
12	Gutmann, T.	Harmonizing corporate venturingmodes: An integrative review and research agenda	2019
13	Lukosiute,K. Jensen,S. Tanev, S.	Is joining a business incubator or accelerator always a good thing?	2019
14	David-West, O. Umukoro, I. Onuoha, R.	Platforms in Sub-Saharan Africa: Startup models and the role of business incubation	2018
15	Springer, L., Michelis, D., Senges, M.	How traditional companies can foster innovation through collaboration with startups	2018
16	Röhm, P.	Exploring the landscape of corporate venture capital: a systematic review of the entrepreneurial and finance literature	2018
17	Selig, J. C. Assir, T. Baltes, H. G.	How corporate accelerators Foster organizational transformation: An internal perspective corporate	2018
18	Mahmoud-Jouini, B. S. Duvert, C. Esquirol, M.	Key factors in building a corporate accelerator capability: Developing an effective corporate accelerator requires close attention to the relationships between startups and the sponsoring company	2018

(continued)

Table 2.1 (continued)

No	Author	Title	Year
19	Kreusel, N., Roth, N., Brem, A.	European business venturing in times of digitisation – An analysis of for-profit business incubators in a triple helix context	2018
20	Jackson, P., Richter, N.	Situational logic: An analysis of open innovation using corporate accelerators	2017
21	Kupp, M. Marval, M. Borchers, P.	Corporate accelerators: Fostering innovation while bringing together startups and large firms	2017
22	Hobbs, G. K. Link, N. A. Scott, T. J.	Science and technology parks: An annotated and analytical literature review	2017
23	Livieratos, A.D., Lepeniotis, P.	Corporate venture capital programs of European electric utilities: Motives, trends, strategies and challenges	2017
24	Kohler, T.	Corporate accelerators: Building bridges between corporations and startups.	2016
25	Wang, X. Edison, H. Bajwa, S. S. Giardino, C. Abrahamsson, P.	Key challenges in software startups across life cycle stages	2016
26	Mocker, V. Bielli, S. Haley, C.	Winning together: A guide to successful corporate-startup collaboration	2015
27	Capdevila, I.	Co-working spaces and the localised dynamics of innovation in Barcelona	2015
28	Salamzadeh, A., Kawamorita Kesim, H.	Startup companies: Life cycle and challenges	2015
29	Weiblen, T. Chesbrough, H	Engaging with startups to enhance corporate innovation	2015
30	Saebi, T.Nicolai J.	Business models for open innovation: Matching heterogenous open innovation strategies with business model dimensions	2014
31	Dempwolf, C. S., Auer, J., and D'Ippolito, M.	Innovation accelerators: Defining characteristics among startup assistance organizations.	2014
32	Von, V. Dauderstädt, P.	Success factors in strategic corporate venturing	2013
33	Hurley, M. hunter, R.	Gateway to growth: Innovation in oil and gas industry	2013
34	Nagji, B., and Tuff, G.	Managing your innovation portfolio	2012
35	Reimsbach, D. Hauschild, B.	Corporate venturing: An extended typology.	2012
36	Minshall, T.H.W. Mortara, L. Napp, J.J. Probert, D.R.	Making "asymmetric" partnerships work	2010
37	Leiponen, A. Helfat, C. E.	Innovation objectives, knowledge sources, and the benefits of breadth	2010
38	Hackett, S. Dilts, D.	Inside the black box of business incubation: Study B – Scale assessment, model refinement, and incubation outcomes	2008

(continued)

Table 2.1 (continued)

No	Author	Title	Year
39	Jenni, M., Frischknecht, P.	Business incubators – An integrated technology transfer strategy	2008
40	Maula, M.	Corporate venture capital as a strategic tool for corporations	2007
41	Heitlager, I., Helms, R., and Brinkkemper, S.	A tentative technique for the study and planning of co-evolution in product in software evolvability	2007
42	Birkinshaw J, Hill S.	Corporate venturing units: Vehicles for strategic success in the new Europe	2005

placed in a wide spectrum of increased engagements which are informal, outsourcing, partnership/alliances and mergers, and acquisition [52]. Gilsing and associates [22] examined the three structural features of alliance network portfolio: position, focus, and density. Faems and associates [20] provided a conceptual framework that investigated the effect of both an increase in value and an increase in the number of technological alliances on the financial performance [20].

3.2 *Constructs of Our Collaboration Framework*

We have extracted the constructs of our framework by analyzing the literature.

Innovation Strategies and Goals of the Established Company For established businesses, past and present successes can stifle forward thinking. Deciding for innovation is a fundamental strategic decision for any company [25]. In a report published by PwC [62], 80% of interviewed companies stated that innovating is important for them, but only half of these companies admitted they had a well-defined strategy for it, and this is a serious problem. Usually, open innovation strategies categorize the insourcing of innovation based on the "breadth" and "depth" of the search for knowledge. Breath of knowledge search [42] is often defined as the number of various types of external partners; and the depth of knowledge search refers to those external partners that are deeply integrated into an innovative activity of the company [42, 67]. Overall, strategic goals and financial goals are the two main types of innovation goals [9, 50]. In various sources, the goals of big companies for partnership with startups have been listed [19, 37, 50, 53, 81]. Varesco Kager et al. [80] presented examples of startup collaboration goals, such as focusing on generating benefits for the company's image and culture of the company and the ecosystem development more generally, as well as testing and tapping into new business areas [80]. The large firm's innovation goals and strategies have a profound effect on its collaboration with startups.

Innovation Portfolio of the Established Company Gutmann and Lang [27] stated: "established companies should search for, acquire and develop new resources and capabilities to go in a new direction; break some rules and promote variance and slack; serve new customers with new needs; develop and lead new competencies." Allmendinger and Berger have stated [1] one reason for a startup to choose a company for partnership over others is that the company's portfolio is similar to their own portfolio. The Innovation Ambition Matrix is one of the innovation frameworks, containing two main dimensions of products and markets [55]. Nagji and Tuff defined three primary types of innovation: core, adjacent, and transformational and recommended that companies should dedicate 70% of their resources to core innovation, 20% to adjacent innovation, and 10% to transformational innovation [55]. A balanced innovation portfolio means that a correct proper mixture of investment on gradual, radical, and breakthrough innovation must be existent in all areas of innovation portfolio [62]. The company's balanced innovation portfolio determines the type of innovation that the company intends to invest in and the degree to which the company wants to focus on that innovation type. This will give an outlook to the company to have a better insight about its desire innovation and accordingly its collaborations with startups.

The Maturity Level of the Corporate Venturing in the Established Company Direct engagement of large organization actors with startups is critical [33]. Companies typically assign innovation managers to manage relationships with startups and the program. Some companies may involve other corporate members who interacted with the startups in the collaboration's operational execution [80]. The term "corporate venturing" has attracted the attention of some researches [4, 26]. Corporate venturing has been adopted from companies like Siemens, GE, IBM, Merck, Lucent, and Xerox [59, 60]. Corporate venturing acts as a strategic tool for companies to generate, acquire, and transfer innovations [26]. As a general concept, it covers all the tools and mechanisms for cooperating with startups and companies can use them according to the situation, strategy, or even the speed of innovation in their industry [59, 60]. Reimsbach and Hauschild [63] present a broad typology in three dimensions: focus of corporate venturing, degree of arbitration in the corporate venturing process, and explorative or exploitative orientation. Prats et al. stated that most companies have various tools at their disposal for cooperating with startups and each company uses the appropriate mechanism depending on the maturity level of the venturing unit in that company [59, 60]. They pointed out some factors should be focused on when deciding on the collaboration mechanisms: profitability, waiting duration for observing the effect of mechanism on the outcomes of the company, intensity of capital spending, engagement and participation level, and which stage of innovation phase is being protected in each mechanism. Prats et al. [61] defined three maturity levels for the corporate venturing unit: the building stage (less than 3 years) the scaling stage (between 4 and 10 years) and the consolidating stage (more than 10 years) [61]. Whether the company has experience of working with startups or what mechanisms of collaboration have been used in the company and how successful previous experiences were helps the company to realize the best way for cooperating with startups.

The Needs and Expectations of Startups Corporates and startups may differ in their expectations, demands, beliefs, and cultures [34, 52, 78]. The large firms in order to avoid mistakes that would jeopardize their success must design collaboration programs with startups in mind [58, 83]. Also, startups should be conscious of their expectations before cooperating with a big company and joining their startup programs such as corporate incubators and accelerators. Linna [44] stated that the interest of startups in cooperating in the literature is convergent. The needs and expectations of startups have been proclaimed by Kohler [37] to include getting access to resources, increasing the credit, getting access to markets, and receiving budget. Lukosiute and associates [46] mentioned some of the negative experiences of startups after joining the corporate incubator and accelerator programs and how these programs were not matching expectations of startups [46]. As startups have varied expectations, a company that intends to cooperate with startups should identify these needs correctly in order to respond to them.

The Life Cycle Stages of Startups Startups are essentially varied and complex, yet they have their own life cycle. Heitlager and associates [30] studied the evolutionary patterns of different software startups, and they believe that all the stages of a startup's evolution are taken consciously and, therefore, these stages can be studied. Salamzadeh and Kawamorita Kesim divided the lifespan of a startup into three stages: bootstrapping, seed, and creation stages. They summarized: the life cycle begins with an idea and ends with an exit strategy such as merger, acquisition, IPO, etc. [69]. Marmer envisioned six stages for each startup to go through: discovery, validation, efficiency, scale, sustain, and conservation [49]. Overall and Wise [56] defined four stages for startup's life cycle: pre-seed, early stage, expansion, and later stage. In each stage of their life cycle, startups need different investment from different stockholders [56]. When it comes to investment, startups can have various supporters in each stage which include: (1) the investor of seed and cultivation of idea stage [43], (2) the angel investor [79], and (3) venture capital [23]. Also, in each startup's stage of life cycle, the risk level of collaboration and the readiness level of their provided MVP are different [68]. Minimum viable product (MVP) of a startup product is a low-quality early sample [35]. Reisdorfer-Leite et al. [64] have determined the stages of startup life cycle from the product life cycle, consisting of three main stages: (1) pre-startup stage which corresponds to beginning of product life (BOL), including design and production; (2) startup stage which corresponds to middle of product life (MOL), including logistics (distribution), use, service, and maintenance; and (3) consolidated enterprise stage which corresponds to end of product life (EOL), including reverse logistics, remanufacturing, reuse, recycle, and disposal. The stage of life cycle can also present the technological readiness level (TRL) of startup's MPV (product) which has some effects on the collaboration [15]. The lower the TRL of the startup is, the higher its uncertainty and risk are for the company. Furthermore, the expectations of the startup may vary, depending on its TRL. Using TRL is also a method for estimating the maturity of technologies in the acquisition and takeover phase [48].

The Proposed Value of a Collaboration Mechanism New methods of collaboration with startups have emerged [59, 60]. More common methods are explained by the following research works: corporate accelerators [13, 27, 40, 47, 71], corporate incubators [29, 39, 54], and venture capital [9, 11, 16, 18, 28, 31, 32, 36, 45, 50, 66, 83]. Non-equity models exist as well, in which the large firm does not take any equity in the startup. Cocreation, colocation, platforms, startup programs, and startup supplier programs are examples of these models [41, 73, 76, 77]. Also, other non-equity models are introduced by researchers, like day-long events [6, 53], technology hub [14]; coworking space [8, 17], and scouting mission [61]. Collaboration models can vary according to several dimensions: duration, location, level of structure, presence of reimbursement for the startup, and if the corporation takes equity or not in startup [76, 77, 83]. When a large corporation decides to collaborate with a startup, it should offer appealing values by means of its collaboration models to be able to absorb the startup. Indeed, the established company should respond to the needs and expectations of startups through various collaboration methods.

Cost and Required Resources for Collaboration The big company must choose its appropriate method among various collaboration mechanisms. Cost, required resources, and duration for each one are different. Innovation competition is short term while incubators may last for more than a few years [16]; and for accelerators duration is generally 1–3 months [10]. Also, the value proposed in each collaboration mechanism is different. The cost required for different mechanism—from lowest to highest—is resources sharing, challenging prizes, hackathons, scouting missions, excubators, venture client, strategic partnership, incubators, accelerators, corporate venture capital, and merger and acquisition [61]. Each mechanism has a different cost and resource, based on a variety of factors such as the values proposed to startups and the average duration of the mechanism.

3.3 The Collaboration Process of an Established Company with Startups

Our proposed funnel-shaped framework for designing the collaboration of an established company with startups is shown in Fig. 2.1.

Authors believe that four steps should respectively be taken by the established company to collaborate with the startups, as follow.

1. Defining and determining the related constructs of the established company, including innovation objectives and strategies, innovation portfolio, and the level of corporate venturing maturity. This step will help the big company to define in detail its innovation goals, demands, research fields, their priorities, and the required time for collaborating with startups.
2. Defining and determining the related constructs of the startups, including expectations and demands of startup and the stages of startup life cycle. This step will help the established company to get to know the startups' expectations, their demands, ideas, MPVs, and their proposed products' TRLs.

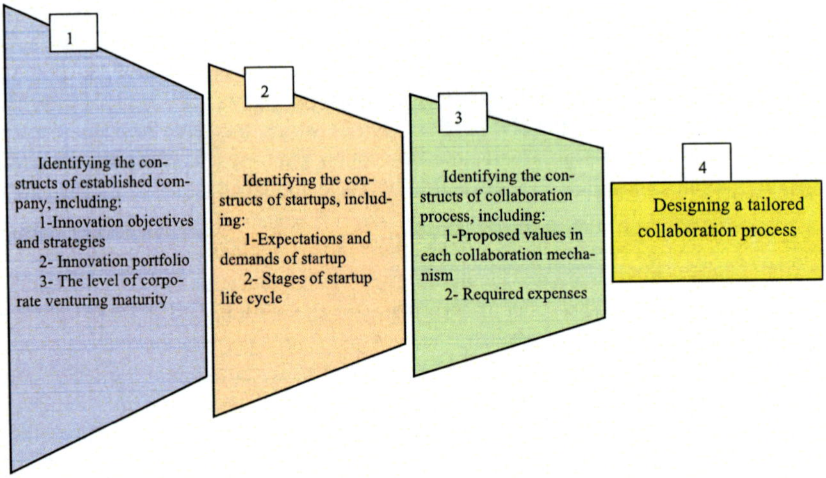

Fig. 2.1 The funnel-shaped framework for designing the collaboration of an established company with startups

3. Defining and determining the related constructs of the collaboration process, including proposed values in each collaboration mechanism and required expenses and resources for each mechanism. This step will help the company to have an insight regarding the required budget, resources, and time to collaborate with startups.
4. Designing a tailored collaboration process. There is a need to develop a portfolio of corporate–startup collaboration initiatives, each of which may necessitate a different setup [74, 75]. This step will help the established company to match its innovation demands, goals, and resources in line with the startup expectations, their life cycle stage, and TRLs and design its effective collaboration mechanisms and models and prepare required procedures and work instructions.

4 Discussion, and Limitations of Research

The results of this article can guide and organize the future researches and prepare the ground for producing accumulative and generalizable findings. These features can become the focus of theoretical and empirical studies of the future on the topic. This article helps the companies and startups that lack the experience of working with each other to plan their collaboration accordingly.

Although our study has achieved considerable results, there are some grounds that probably limit its implications. Considering the systematic analysis of this paper's subject literature, some limitations were bound to arise since some biased decisions that may have included or excluded a paper could not have been avoided. Additionally, the nature of qualitative researches gives rise to unavoidable interpretations by the researcher.

5 Future Studies

More studies are required to test and examine our framework. Also, more researches are needed to analyze the effects of politics, industry, and economic condition on the configuration of our framework. The third area for future researches can be concerned with defining the criteria for a successful collaboration, based on the constructs of our framework.

6 Conclusion

The idea behind this article is that a company can identify its related specific criteria and then carefully screen and select potential startups using determined factors from among the many factors that are effective for cooperation with startups. Then, the subprocesses of collaboration with startups can then be adjusted based on the analysis of these defined factors.

We proceeded to review the literature concerning collaboration between an established company and startups and attempted to move towards organizing this scientific body of knowledge. We have divided the literature review into three dimensions of startup, big company, and collaboration process. We also identified the influential constructs of the collaboration, based on the prior researches, including (1) innovation objectives and strategies, (2) innovation portfolio, (3) the level of corporate venturing maturity, (4) expectations and demands of startup, (5) stages of startup life cycle, (6) proposed values in each collaboration mechanism, and (7) required expenses. We have built a framework which explains four steps should be taken to have a successful collaboration. We have presented the process of collaboration and its four steps based on our framework.

References

1. Allmendinger, M., & Berger, E. (2019). Selecting corporate firms for collaborative innovation: Entrepreneurial decision making in asymmetric partnerships. *International Journal of Innovation Management.*, World Scientific Publishing Co. Pte. Ltd, *24*(01), 1–34.
2. Avnimelech, G., & Teubal, M. (2006). Creating venture capital industries that co-evolve with high tech: Insights from an extended industry life cycle perspective of the Israeli experience. *Research Policy, 35*(10), 1477–1498.
3. Bastié, F., Cieply, S., & Cussy, P. (2013). The entrepreneur's mode of entry: The effect of social and financial capital. *Small Business Economics, 40*(4), 865–877.
4. Birkinshaw, J., & Hill, S. (2005). Corporate venturing units: Vehicles for strategic success in the new Europe. *Organizational Dynamics, 34*(3), 247–257.
5. Blank, S. (2010). What's a startup? *First Principles, 1*, 25. https://steveblank.com/2010/01/25/whats-a-startup-first-principles/

6. Briscoe, G., & Mulligan, C. (2014). *Digital innovation: The hackathon phenomenon*. Reativeworks London. Working Paper No. 6, May 2014, Retrieved Mar. 12, 2019.
7. Campbell, A., Birkinshaw, J., Morrison, A., & Batenburg, R. V. B. (2003). The future of corporate venturing. *MIT Sloan Management Review, 45*(1), 30–37.
8. Capdevila, I. (2015). Co-working spaces and the localised dynamics of innovation in Barcelona. *International Journal of Innovation Management, 19*(03).
9. Chesbrough, H. (2003). *Open Innovation* (Vol. 45, p. 33). Harvard University Press.
10. Cohen, S., & Hochberg, Y. V. (2014). Accelerating startups: The seed accelerator phenomenon. *SSRN Electronic Journal, 2014*, 1–16. https://doi.org/10.2139/ssrn.2418000
11. Cordes, A., Guderian, C. C., & Riar, F. J. (2021). Exploring the practice of evaluation in corporate venturing. *International Journal of Innovation and Technology Management, 18*(5).
12. Corvello, V., Steiber, A., & Alänge, S. (2021). Antecedents, processes and outcomes of collaboration between corporates and start-ups. *Review of Managerial Science, 17*, 129.
13. Crişan, E. L., Salanță, I. I., Beleiu, I. N., Bordean, O. N., & Bunduchi, R. (2021). A systematic literature review on accelerators. *The Journal of Technology Transfer, 46*, 62–89.
14. David-West, O., Umukoro, I., & Onuoha, R. (2018). Platforms in Sub-Saharan Africa: Startup models and the role of business incubation. *Journal of Intellectual Capital, 19*(3), 581–616.
15. De La Tour, A., Soussan, P., Harlé, N., & Chevalier, R. (2017). *Fostering collaboration between corporates and startups*. Boston Consulting Group, and Hello Tomorrow.
16. Dempwolf, C. S., Auer, J., & D'Ippolito, M. (2014). *Innovation accelerators: Defining characteristics among startup assistance organizations*. SBA Office of Advocacy.
17. Du Boucher, V. (2016). *GSMA A few things we learned about tech hubs in Africa and Asia |* Mobile for Development. August 5. www.gsma.com/mobilefordevelopment/programme/ecosystem-accelerator/things-learned-tech-hubs-africa-asia/
18. Dushnitsky, G. (2009). Corporate venture capital: Past evidence and future directions. In G. Dushnitsky (Ed.), *The Oxford Handbook of Entrepreneurship*. Oxford University Press.
19. Dushnitsky, G., Haemming, M., & Tharp, D. (2008). *Global corporate venture capital survey 2008–09: Benchmarking programs & practices*. Ernst & Young.
20. Faems, D., De Visser, M., Andries, P., & Van Looy, B. (2010). Technology alliance portfolios and financial performance: Value-enhancing and cost-increasing effects of open innovation. *Journal of Product Innovation Management, 27*(6), 785–796.
21. Fortnum, D., Mead, W., Pollari, I., Hughes, B., & Spei. (2016). *The Pulse of Fintech Q4 2016: Global analysis of investment in Fintech. 2*, 3. https://www.insead.edu/news/2016-insead-and-500-startups-report
22. Gilsing, V., Nooteboom, B., Vanhaverbeke, W., Duysters, G., & Van Den Oord, A. (2008). Network embeddedness and the exploration of novel technologies: Technological distance, betweenness centrality and density. *Research Policy: A Journal devoted to Research Policy, 37*(10), 1717.
23. Gompers, P. A. (1994). The rise and fall of venture capital. *Business and Economic History, 23*(2), 1–26.
24. Graham, P. (2012). *Startup is growth*. http://www.paulgraham.com/growth.html.
25. Gunday, G., Ulusoy, G., Kilic, K., & Alpkan, L. (2011). Effects of innovation types on firm performance. *International Journal of Production Economics, 133*(2), 662–676.
26. Gutmann, T. (2019). Harmonizing corporate venturingmodes: An integrative review and research agenda. *Management Review Quarterly, 69*, 121–157.
27. Gutmann, T., & Lang, C. (2022). Unlocking the magic of corporate-startup collaboration: How to make it work. *IEEE Engineering Management Review (IEEE), 50*(2), 19–25.
28. Hackett, S., & Dilts, D. (2008). Inside the black box of business incubation: Study B – Scale assessment, model refinement, and incubation outcomes. *Journal of Technology Transfer, 33*(5), 439–471.
29. Hausberg, J. P., & Korreck, S. (2020). Business incubators and accelerators: a co-citation analysis-based, systematic literature review. *The Journal of Technology Transfer, 45*, 151–176.
30. Heitlager, I., Helms, R., & Brinkkemper, S. (2007). *A tentative technique for the study and planning of co-evolution in product software startups* (pp. 42–47). IEEE.

31. Hobbs, K. G., Link, A. N., & Scott, J. T. (2017). Science and technology parks: an annotated and analytical literature review. *Journal of Technology Transfer, 42*, 957–976.
32. Huang, P., & Madhavan, R. (2020). Dumb money or smart money? Meta-analytically unpacking corporate venture capital. *Strategic Entrepreneurship Journal, 15*(3), 403.
33. Hübel, C., Weissbrod, I., & Schaltegger, S. (2022). Strategic alliances for corporate sustainability innovation: The 'how' and 'when' of learning processes. *Long Range Planning, 55*, 102200.
34. Jackson, P., & Ritcher, N. (2017). Situational logic: An analysis of open innovation using corporate accelerators. *International Journal of Innovation Management, 21*(07), 1750062.
35. Jawanda, P. (2015). *The lean start-up approach versus scrum – A case study of German startup StudentCouch*. University of Twente.
36. Jenni, M., & Frischknecht, P. (2008). Business incubators – An integrated technology transfer strategy. *EMPA Activities*, 95–96.
37. Kohler, T. (2016). Corporate accelerators: Building bridges between corporations and startups. *Business Horizons, 59*(3), 347–357.
38. Kopera, S., Wszendybył-Skulska, E., Cebulak, J., & Grabowski, S. (2018). Interdisciplinarity in tech startups development – Case study of 'Unistartapp' project. *Foundations of Management, 10*(1), 23.
39. Kreusel, N., Roth, N., & Brem, A. (2018). European business venturing in times of digitisation – An analysis of for-profit business incubators in a triple helix context. *International Journal of Technology Management, 76*(1–2), 104–136.
40. Kupp, M., Marval, M., & Borchers, P. (2017). Corporate accelerators: Fostering innovation while bringing together startups and large firms. *Journal of Business Strategy, 38*(6), 47–53.
41. Kurpjuweit, S., & Wagner, S. M. (2020). Startup supplier programs: A new model for managing corporate-startup partnerships. *California Management Review, 62*(3), 64.
42. Leiponen, A., & Helfat, C. E. (2012). Innovation objectives, knowledge sources, and the benefits of breadth. *Strategic Management Journal, 31*(2), 224–236.
43. Lesonsky, R. (2007). *Start your own business*. Entrepreneur Press.
44. Linna, E. (2017). Corporate startup growth program (Master Theses). Aalto University, Finland.
45. Livieratos, A. D., & Lepeniotis, P. (2017). Corporate venture capital programs of European electric utilities: Motives, trends, strategies and challenges. *Electricity Journal, 30*(2), 30–40.
46. Lukosiute, K., Jensen, S., & Tanev, S. (2019). Is joining a business incubator or accelerator always a good thing? *Technology Innovation Management Review, 7*(10), 5–15.
47. Mahmoud-Jouini, B. S., Duvert, C., & Esquirol, M. (2018). Key factors in building a corporate accelerator capability: Developing an effective corporate accelerator requires close attention to the relationships between startups and the sponsoring company. *Research Technology Management, 61*(4), 26–34.
48. Mankins, J. C. (2009). Technology readiness assessments: A retrospective. *Acta Astronautica, 65*, 1216–1223.
49. Marmer, M., B. L. Herrmann, E. Dogrultan, R. Berman, C. Eesley, and S. Blank. Startup genome report extra on premature scaling. Startup Genome, 2011.
50. Maula, M. (2007). Corporate venture capital as a strategic tool for corporations. In *Handbook of research on venture capital* (Vol. 1, pp. 371–392). Edward Elgar Publishing.
51. McClure, D. (2013). *What is the proper definition of a startup*. https://www.quora.com/Entrepreneurship/What-is-the-proper-definition-of-a-startup/answer/Dave-McClure
52. Minshall, T. H. W., Mortara, L., Napp, J. J., & Probert, D. R. (2010). Making "asymmetric" partnerships work. *Research Technology Management, 53*(3), 53–63.
53. Mocker, V., Bielli, S., & Haley, C. (2015). *Winning together: A guide to successful corporate-startup collaboration*. Nesta.
54. Morgan, H. (2014). Venture capital firms and incubators. *Research-Technology Management, 57*(6), 40–45.
55. Nagji, B., & Tuff, G.. (2012). Managing your innovation portfolio. *Harvard Business Review*.
56. Overall, J., & Wise, S. (2015). An S-curve model of the start-up life cycle through the lens of customer development. *The Journal of Private Equity, 18*(2), 23–34.

57. Paternoster, N., Giardino, C., Unterkalmsteiner, M., Gorschek, T., & Abrahamsson, P. (2014). Software development in startup companies: A systematic mapping study. *Information and Software Technology, 56*(10), 1200–1218.
58. Prashantham, S., & Kumar, K. (2019). Engaging with startups: MNC perspectives, *IIMB Management Review, 31, 4*, 407–417.
59. Prats, M. J., Canonici, T., Siota, J., & Contijoth, X. (2018). *Open innovation, building, scaling and consolidating your firm's corporate venturing unit*. IESE Business School and Opinno.
60. Prats, M. J., Siota, J., Canonici, T., & Contijoch, X. (2018). *Corporate venturing: A David-and-Goliath collaboration*. IESE.
61. Prats, M. J., Siota, J., Martinez-Monche, I., & Martínez, Y. (2019). *Open innovation: Increasing your corporate venturing speed while reducing the cost*. IESE Business School University of Navarra..
62. PwC. (2013). *Gateway to growth: innovation in the oil and gas industry*. PwC.
63. Reimsbach, D., & Hauschild, B. (2012). Corporate venturing: An extended typology. *Journal of Management Control : Zeitschrift für Planung & Unternehmenssteuerung, 23*(1), 71–80.
64. Reisdorfer-Leite, B., de Oliveira, M. M., Rudek, M., Szejka, A. L., & Junior, O. C. (2020). Startup definition proposal using product lifecycle management. In *IFIP international conference on product lifecycle management* (pp. 426–435).
65. Ries, E. (2011). *The lean startup: How today's entrepreneurs use continuous innovation to create radically successful businesses*. Crown Business.
66. Röhm, P. (2018). Exploring the landscape of corporate venture capital: A systematic review of the entrepreneurial and finance literature. *Management Review Quarterly, 68*(3), 279–319.
67. Saebi, T., & Nicolai, J. (2015). Business models for open innovation: Matching heterogeneous open innovation strategies with business model dimensions. *European Management Journal, 33*(3), 201–213.
68. Salamzadeh, A. (2015). New venture creation: Controversial perspectives and theories. *Economic Analysis, 48*(3–4), 101–109.
69. Salamzadeh, A., & Kawamorita Kesim, H. (2015). Startup companies: Lifecyle and challenges. In *4th international conference on employment, education and entrepreneurship (EEE)*. Belgrade, Serbia.
70. Salamzadeh, A., & Kawamorita Kesim, H. (2017). The enterprising communities and startup ecosystem in Iran. *Journal of Enterprising Communities, 11*(4), 456–479.
71. Selig, J. C., Assir, T., & Baltes, H. G. (2018). How corporate accelerators foster organizational transformation: An internal perspective corporate. *IEEE international conference on engineering, technology and innovation (ICE/ITMC)*.
72. Sommer, S. C., Loch, C. H., & Dong, J. (2009). Managing complexity and unforeseeable uncertainty in startup companies: An empirical study. *Organization Science, 20*(1), 118–133.
73. Srinivasan, A., & Venkatraman, N. (2018). Entrepreneurship in digital platforms: A network-centric view. *Steretegic Entrepreneurship Journal, 12*(3). https://doi.org/10.1002/sej.1272
74. Steiber, A., & Alänge, S. (2020). Corporate-startup co-creation for increased innovation and societal change. *Triple Helix Journal*, 1. https://doi.org/10.1163/21971927-bja10004
75. Steiber, A., & Alänge, S. (2020). Corporate-startup collaboration: Effects on large firms' business transformation. *European Journal of Innovation Management, 24*(2), 235–257.
76. Steiber, A., Alänge, S., & Corvello, V. (2021). Evaluating corporate-startup co-creation: A critical review of the literature. *International Journal of Innovation Management, 25*(07).
77. Steiber, A., Alänge, S., & Corvello, V. (2021). Learning with startups: An empirically grounded typology. *The Learning Organization, 28*(2), 153–166.
78. Urbaniec, M., & Agnieszka, Ż. (2021). Business model innovation in corporate entrepreneurship: Exploratory insights from corporate accelerators. *International Entrepreneurship and Management Journal, 17*, 865–888.
79. Van Osnabrugge, M., & Robinson, R. J. (2000). *Angel investing: Matching startup funds with startup companies*. Wiley.

80. Varesco Kager, N., Sparr, J. L., & Grote, G. (2022). Looking for guidance? Five principles for leveraging tensions in corporate–startup collaboration. *The Journal of Applied Behavioral Science, 58*(4). https://doi.org/10.1177/00218863221117850
81. Von, V., & Dauderstädt, P. (2013). *Success factors in strategic corporate venturing*. Universität Berlin.
82. Wang, X., Edison, H., Bawja, S. S., Giardino, C., & Abrahamsson, P. (2016). Key challenges in software startups across life cycle stages. *Lecture Notes in Business Information Processing, 251*, 169–182.
83. Weiblen, T., & Chesbrough, H. (2015). Engaging with startups to enhance corporate innovation. *California Management Review (University of California, Berkeley), 57*(2), 66–90.

Chapter 3
Digital Well-Being and Satisfaction of University Students with Online Education

Ingrid Sekelová, Peter Korba, Martina Koščáková, and Samer Al-Rabeei

1 Introduction

Digital technologies are becoming more and more involved in our everyday lives. With digital requirements at the workplace, educational institutions have to adapt their curricula and implement the teaching of skills that were not even known just a few years ago. Digital literacy is a must-know for a young graduate that is looking for a job in the modern world. Advantages come together with several disadvantages in terms of omnipresent digital technology.

Even before the pandemic, the online study started to increase in popularity, directly or indirectly. The emergence of the pandemic only strengthened the position of digital technologies in online education. The use of such new digital technologies has raised many questions in relation to the impact of computer technology, software, and technological advancements. Student satisfaction with online education through digital technologies depends on the instructor, course design, ICT orientation, conscientiousness, open-mindedness, and overall agreeableness [1].

The influence on student satisfaction can be also made by external variables. One of these variables is why people generally adopt or reject the use of certain technology. To better illustrate this decision process of individuals, a proposed technology acceptance model (TAM) was developed to explain this phenomenon [2]. This model is used to evaluate whether an information system will be adopted or rejected by the users. The use of IT in teaching is upheld by the ability to encourage innovation and its ease of use. Through IT operations, it is possible to create new

I. Sekelová (✉) · P. Korba · M. Koščáková · S. Al-Rabeei
Technical University of Košice, Faculty of Aeronautics, Košice, Slovakia
e-mail: ingrid.sekelova@tuke.sk; peter.korba@tuke.sk; martina.koscakova@tuke.sk; samer.al-rabeei@tuke.sk

L. Knapcikova, S. Zohrehvandi (eds.), *9th International Conference on Mobility, IoT and Smart Cities*, EAI/Springer Innovations in Communication and Computing,
https://doi.org/10.1007/978-3-031-28225-6_3

learning spaces, facilitate the transformation of teaching activities, and provide new learning opportunities [3, 4]. According to the technology acceptance model (TAM), the learning outcomes of students are influenced by the effectiveness of the used technology [5]. Educators have a crucial position in the education process and as well in the use of modern digital technologies for education.

However, the implementation of new technologies in the education is a gradual process and takes time. Some countries progress faster than others, but all had to promptly adapt to the unprecedented situation created by the pandemic COVID-19, which promoted the vast use of online platforms. Many education institutions had no other option than to employ new concepts in their teaching to ensure continuous education even in the time of lockdown. The pandemic forced educational institutions to move toward increased learning of digital literacy not only among their students but also employees-educators. Suddenly, there was a new need for skills that were not a priority before and had to be mastered in a short time often by self-study.

The development of digitalization in education is on the rise in the form of various e-learning programs and the use of artificial intelligence or virtual reality. Some universities have been providing online courses that are accessible from home, work, or the university library. Such vast options also promote the internationalization of education and its accessibility as it is possible to connect to such course from anywhere in the world. The use of virtual reality may help in the practical training for various professions, such as pilots, by making the experience more realistic.

While preparing a course that is to be taught through online tools, attention should be certainly paid to the course design, which has an influence on student satisfaction. There are several parts to the design of an e-learning course, which include the structure, the interface, testing and evaluation methods, and a forum for lecturers and learners to exchange ideas [6]. Many e-learning courses contain chapters of learning materials that can be used as additional materials that will help students gain a deeper understanding of the subject matter [7].

A study performed in 2021 [8] subjected around 400 students to an interview process with a goal to evaluate their online learning outcomes during the COVID-19 pandemic. According to the findings, it was revealed that the following factors have an impact on the students' outcomes in descending order are learner characteristics, perceived utility, course content, course design, convenience of use, and faculty capacity.

In order to assess the satisfaction of university students with online teaching, this article focuses on the concept of digital well-being and its position in the education process. It presents a study addressing the satisfaction of students of the Faculty of Aeronautics at the Technical University of Košice with the progression in teaching procedures during the pandemic. It compares the results from an earlier study with the present study and tracks whether there is a shift to greater satisfaction in terms of the use of online platforms. Analysis of the questionnaires permits the authors to provide recommendations and suggestions for possible improvements that can be made in terms of teaching through online platforms.

2 Digital Well-Being

The progressively advancing technology opens up the issue of the digital well-being of humans, which is understood as living a life that is good for an individual even with the influence of digital technologies in an informational society [9]. The perception of well-being can be viewed as subjective, and it generally focuses on three aspects of a person's life: relationships, work, and health. In terms of digital well-being, the focus is centered on how digital technologies influence the three abovementioned aspects [10]. Nowadays, the well-being of a human is linked to the digital environment with which they come into contact. The questions about the influence of digital technology on the well-being of people became even more pressing with the emergence of the pandemic of COVID-19.

Generally, the presence of technology should improve the quality of life of people. There are numerous benefits that modern technology brings in the form of higher productivity or reduced social inequality as more people have access to healthcare or various information [11]. However, positive impacts are opposed to concerns about the mental health of individuals [12], which may be attributed to the excessive use of technologies or the fear of continuing automation in the labor market [13]. Up-to-day education has to reflect the needs of the workplace which requires a new set of skills that were not present before in the form of digital literacy. Therefore, it is inevitable for students to apprehend these skills at their earliest convenience as employees are required to have these skills and continuously adapt to the mastery of similar ones [14].

One of the possibilities of how to cope with the overpresence of digital technology in our lives is to temporarily disconnect from digital media; even though the effectiveness of such disconnect remains uncertain, further research may address the effects of self-determined media habits [10, 15]. However, in certain situations, various duties from work or school may forbid us from disconnecting. Negative impacts of digital use may cause depression, Internet or smartphone addiction, and social comparison. Conscious use of digital technologies for accessing quality information, communication, or entertainment assumes positive effects on individuals' lives.

While digital technology is becoming omnipresent and, in many cases, unavoidable, it is not accessible to the whole population for several reasons. A successful operation of digital technology requires a certain level of digital literacy which may pose problems for example, for elderly people or the ones who are not as skilled in the use of digital technologies. Further, the purchase price of digital technologies is still quite high which influences the low-income population that is not able to keep up with purchasing specific devices [16]. These issues became more apparent during the pandemic of COVID-19 in which special attention was paid to social distancing and thus many basic aspects of life had to be performed without personal contact and through digital technologies.

The emergence of the pandemic was closely related to the integration of digital technologies into daily routines at a rapid rate. Individuals and society as a whole turned to digital technologies for shopping, seeking medical attention, socializing, attending school, or working. In just a few weeks, society changed its way of behaving, and it is probable that even after the end of the pandemic, the behavior will not go entirely to the way it was previously. The pandemic gave a new dimension to the digitalization process; it also augmented the urgency to address and care for our digital well-being [17]. Nowadays, it is not only about the fact whether an individual has an Internet connection and a device that can connect to it, but it is mostly about how stable and reliable those devices and the Internet connection itself are, as the requirements for videoconferencing, which was a very often used during the pandemic, are much higher than the requirements for simple Internet browsing.

The issue of digital well-being during the COVID-19 pandemic was addressed by the Center for Humane Technology which presented guidelines in the form of rules to mitigate the risks coming from managing many of the aspects of life online [18]. By adopting these rules [17], concluded that there are several strategies to cope with digital overwhelm, but to be effective, these strategies are to be implemented collectively. Strategies include limiting our contact with digital technology by preparing a time management plan, having a skeptical attitude to online products, and adopting a mindful approach to determine how digital technologies affect the well-being of an individual.

3 Education During the Pandemic

During the pandemic, the possibilities for education were severely limited. Schools or universities without previous entirely online teaching systems were trying to cope with the situation of closed gates at their institutions the best way they were able to. Some educational institutions had some form of online communication with students had already established, for example, in the form of school emails. However, for some other institutions mostly dedicated to the education of younger students, it was even harder to transfer to the system of online education.

The transformation to online education due to the pandemic deepened the digital divide that was already present. In reaction to the closing of educational institutions, the educational process continued via online means, which made vulnerable students even more disadvantaged. The response to the pandemic depended on the preparedness of the educational systems of each country. Some countries' education systems had already embedded technology and the Internet so they were able to provide adequate working conditions for teachers; however, the nature of the response was likely to show the inequalities among students and other weaknesses in the education system [19]. According to a study [20] in the United States, students who had issues with the technical maintenance of their devices used for the online education process or had unstable Internet had more trouble with keeping up with the deadlines, assigned work, or following online classes. This study revealed that in order to attain success in their studies, students need to have a good high-speed Internet connection and a functional digital device and also a strong human connection with the instructor of the course.

The pandemic COVID-19 caught educational institutions unprepared as in general, the form of online teaching has not been used at universities before, and even if it was, it certainly has not been used to such an extent as it had to be during the lockdown. The first closing of schools and universities in Slovakia was on 12 March 2020 [21], and in the upcoming months, the method of teaching depended on the actual situation and restrictions posed by the government. After the first shock which everyone experienced, the Student Council for Higher Education provided some recommendations [22], together with the European Union [23] on how to maintain the quality of education as high as possible even in a world intertwined with the pandemic. Despite the recommendations, online teaching has deprived students of opportunities to network and socialize with classmates and teachers, which severely affected their psychological well-being [24].

In the spring of 2020, the first closure of universities was believed to be short and temporary in terms of a few weeks. This belief was reflected in the attitudes of the institutions and educators who did not immediately receive recommendations for online teaching. A national study was concluded in March 2020 to address the issues of online learning at universities and the methods that were used by educators. The report from the Student Council for Higher Education highlighted that universities have to enable students to finish the current summer semester by moving the practical sections into the following semester of the next academic year [22]. However, educational institutions were closed due to the pandemic even in the following winter semester which resulted in replacing some elements of practical work with videos, which unfortunately cannot be considered a sufficient substitute for learning skills through direct experience [25]. The study further focused on the methods used for online teaching. The prevailing methods were of passive nature, in which the student is not in direct interaction with the educator such as studying recommended literature, sharing of the presentation created by the educator, writing essays, and working with e-learning materials. The report recommends higher integration of interactive methods in the forms of videoconferences or at least presentations with audio commentary of the educator. A study revealed that only 37% of students are sufficiently informed in order to successfully manage online learning. In addition to unstable and sometimes missing communication, excessive workload and the use of different digital tools by individual educators constitute significant issues for the students. Most of the students were not satisfied with online teaching in their study program, and at the same time, majority did not agree with the statement that online learning fully replaces face-to-face forms of teaching [22].

Since this study featured only approximately 12% of students from the Technical University of Košice, this institution decided to perform its own study aiming at determining the actual state of online teaching at this university. According to a study published in June 2020 at the Technical University of Košice [26], the majority of methods used in online teaching at that time were of passive nature. The ratio of the most common methods used was comparable to the ones from the previous national study, which featured several Slovak universities. Students in their responses stated that many educators make little or no use of any interactive teaching method. In addition to the above, they also mentioned other problems such as the absence of communication, long time intervals in e-mail communication, absence of assessment of submitted assignments, a large number of essays due in a short time interval, and difficulties in gathering all necessary information as each educator updates information in a different place.

The study done by the Technical Universities focused on all faculties including the Faculty of Aeronautics. The present study addresses only the Faculty of Aeronautics as it presented only 4,9% of all respondents in the previous study. The aim is to observe and compare the methods of teaching at this particular faculty approximately a year and a half after the two studies mentioned above. After such time, educators had more time to adapt to the requirement of online teaching and to integrate the various recommendations.

4 Online Teaching Platforms Used at the Faculty of Aeronautics

The unexpected pandemic of COVID-19 created an unprecedented situation that revealed the lack of preparedness in terms of digital equipment necessary for online teaching. The beginning of online teaching was missing guidelines and technological preparedness of educational institutions, which the educators had to promptly cope with. Firstly, they employed various passive forms of online teaching while recommendations state that a more interactive approach should be used. One of the interactive forms of online teaching is done through the use of online platforms that provide support for videoconferences, messaging, and other forms of sharing information. There are various types of platforms each offering certain advantages. Our study focused on the use of two online platforms that are used at the Faculty of Aeronautics, which are Webex Meetings and MS Teams. We are going to compare these two platforms while also mentioning Webex Training, which carries some additional features missing in Webex Meetings. These platforms provide the essential features (camera, microphone, sharing of screen, whiteboard) that an educator and student need in order to hold and actively participate in a videoconference.

4.1 Webex Meetings

Webex Meetings is an online platform developed by Cisco [27, 28] through which it is possible to hold a videoconference with all features which are indispensable (camera, microphone, screen sharing). Webex Meetings provides two ways how to manage videoconferences with students. The first option which requires less time in the first stage of setting up the videoconference is through the personal link of the educator. The other option is that the teacher can create a space (or team) for each class with the students who are supposed to attend that class. Both these options have their advantages.

Upon registration, each account is assigned a unique link, which provides access to a virtual space called a personal room. The educator can share the link to their personal room with their students, who will connect to this link in the time of the class (determined by the university schedule). When connecting through the link of the personal room, there is no need to create spaces, which may seem easier for the teacher in the short run. However, it is not possible to share content such as documents or other files through the personal room. Such sharing of files can be then done through email or another platform. This procedure may be harder to keep track of as students cannot access all information (meaning online classes and documents for classes) in one place. The option of a chat can be a useful feature in times of technological issues, e.g., not properly working microphone. However, the chat that is accessible to the students during the videoconference or files shared during the call is not accessible after the call ends (unless it was saved during the call) [27, 28].

On the other hand, creating spaces for specific classes will take more time at the beginning of the academic year, as adding all students can be time-consuming. However, when sharing documents with students through the created spaces, they have access to it all the time and they can be all found in one place. With the use of spaces, there is no need to send emails with the link to the personal room or additional information about the class as all can be done through the space. The advantage of spaces when compared to a personal room is that when the videoconference is made through the space, files, and chat are accessible at all times [27, 28].

Webex Meetings, whether used in form of spaces or personal rooms, provides a network of students and educators. It is possible to write messages to specific people without the use of an email or schedule a video call with them directly. Such a feature can be useful in group projects where students can communicate in groups or pairs without the use of additional platforms, such as email, Facebook, or WhatsApp. A downside of Webex Meetings is that there is no option to create tests or assignments [27, 28], which is a well-needed feature for educators.

4.2 Webex Training

Even though the feature of testing is not possible in Webex Meeting, it is accessible through Webex Training. To access to the Webex Training is provided from the browser, not from the desktop app Webex. This may seem slightly complicated as not all functions are found at the same place and it may make the creation of tests less accessible. After signing into Webex in the browser, in the menu, we can see the button for Webex Training. After clicking on this a new window will open with Webex Training, which has a different design when compared to the main Webex app, this may seem confusing for some people as they were already accustomed to the design in the app [29, 30].

It is possible to create tests through Test Library. This platform provides several types of questions (essay, multiple choice, fill in the blanks, etc.), and the creation of the test is quite intuitive. Unfortunately, it is not possible to assign the test during a regular videoconference done through a personal room or space. In order to assign the test, there is a need to schedule the meeting through Webex Training. Creating a meeting through Webex Training may seem a little bit complicated; in order to start it, it is necessary to install an extension to the browser. Based on the type of question, e.g., multiple choice, it is possible for the platform to correct the answers itself after setting the correct answers to the questions before the start of the test. The answers to the questions in the form of essays have to be checked by the educator. To evaluate the tests, it is necessary to go back to the Webex Meeting app and find the meeting during which the test was written. Since it is not possible to find all information on one platform but there is the need to switch from one to the other in the browser, it might seem a bit complicated and confusing for people with lower digital literacy [29, 30].

4.3 MS Teams

MS Teams has a special option for education purposes, which gives teachers the possibility to create teams for each class, where they can add all students that should attend the class, similar to the spaces that can be created in Webex Meetings. This procedure of creating teams may seem to require a lot of time and effort, but in the long run, it can be very advantageous. It provides the possibility to share files and to write in the chat, before, during, and after the call, while all this information is always accessible. MS Teams permits the users to chat with specific people and call them directly; however, it does not have the feature of a personal room. This means that if the teacher wants to hold a videoconference with their students, they have to create teams based on the classes.

One of the biggest advantages of MS Teams in the education edition is the option of assignments and tests. It is possible to create assignments (e.g., homework), and it is very easy to keep track of which students have already turned in their assignments, which did not, and which turned them in after the deadline. The educator can check the assignments, give written evaluations, or ask the student to add or correct certain parts of their work. After students receive their evaluation, they can correct their work and turn it in again. This feature permits students to get feedback for their work which is crucial in online learning. It also allows the educator to stay on top of the assignments and clearly see whether there is a student who forgot to turn in their work. Without this feature, it is possible to send an assignment to the teacher via email, which may get lost or overlooked among lots of emails from other students.

The option of quiz or test is very intuitive, even though there are fewer types of questions available as in Webex Training; it is in a more familiar environment since the quizzes are done through Microsoft Forms (comparable to Google Forms). As well as in Webex Training, it is possible to pre-set a correct answer in, e.g., multiple choice type of questions. This allows the platform to correct the quiz automatically, which saves time for the educator and students to receive earlier their evaluation. Naturally, the open-ended questions have to be checked by the educator. If the test combines more types of questions, the platform automatically evaluates some of the questions, and the others are left to be checked by the educator. To grade the test, the educator has to go to the individual team to which the quiz was assigned. In MS Teams, it is possible to access many features in one place, from the desktop app; however, it does not provide the option for personal space [31].

4.4 Comparison of Online Teaching Platforms

In Table 3.1, we can see a comparative analysis of the three platforms and the ways of holding a videoconference they provide. In the left column, different features of these platforms were compared. The feature of a personal room is viewed as very useful, as it is a universal link that can be used repeatedly with numerous participants; therefore, it may seem like a disadvantage that MS Teams and Webex Training do not provide such a function. Creating specific groups of students that attend the same class each week seems practical from the long-term perspective, this feature is accessible on all platforms, but it is not used when conducting videoconferences through the personal room. All ways of conducting videoconferences allow for a chat during the call, and Webex Meetings through spaces and MS Teams provide the access to the history of the chat at all times. This feature can be considered significant for students who can access the information and files that were shared during the class later, for example, when they are going to study for the test.

Storage of documents or other shared files such as presentations or notes seems very useful to the students. Webex Personal link does not accommodate such a function, meaning that the files can be sent during the call into the chat, which does

Table 3.1 Comparative analysis of online teaching platforms [27–31]

	Webex Meetings (personal room)	Webex Meetings (space)	Webex Training	MS Teams
Personal room	✔	✖	✖	✖
Creating teams	✖	✔	✖	✔
Chat during call	✔	✔	✔	✔
Chat before and after the call	✖	✔	✖	✔
Storage of documents and files	✖	✔	✖	✔
Assignments	✖	✖	✖	✔
Tests/quizzes	✖	✖	✔	✔

not have to be accessible after the end of the call or through email or another platform which makes managing classes slightly complicated. Webex Meetings with spaces and MS Teams provide the possibility of always accessible sharing of files.

The feature of assignments is accessible only through MS Teams in the educational version of the platform. It allows the educator to give feedback to students and to keep track of the turned in assignments. Testing in online teaching is crucial; however, it is provided only by two platforms, MS Teams and Webex Training. The advantage of Webex Training is the possibility to form tests, but at the same time, it does not have the functionalities which are featured by spaces in Webex Meetings or teams in MS Teams; however, it gives the possibility to create recurring training sessions.

For educational purposes, MS Teams might be viewed as the best option thanks to the test, assignment, and grade system, whereas Webex Meetings might seem as a great tool for bigger conferences or meetings, which could be done through Webex Meetings' Personal room. When the host does not know the specific number and e-mail addresses of participants, it would be very advantageous to provide the participants only with the link to the host's personal room. When teaching a class with the same set of students every week in Webex Meetings, it would be more convenient to create a space with them, while conducting of test could be done either by Webex Training or on some other platform such as Moodle.

5 Methodology

The aim of the performed survey was to access the situation of online teaching at the Faculty of Aeronautics and to find out to which extent are the students satisfied with such teaching and compare the results with the study conducted in March 2020 by the Technical University of Košice. We focused on the platforms which are used for online teaching since several platforms are used simultaneously. The goal is to determine whether the students are content with the platform that is used the most in their classes and whether they would prefer if all their classes were done through the same platform. Additionally, the results from the present study are compared to the findings gathered in 2020 by the Technical University of Košice. The authors believe that improvement is to be observed regarding the accessibility of information to the students and their satisfaction with online teaching. It is expected that there will be also an increase in the percentage of students who view online teaching as fully equivalent to face-to-face teaching since the methods of teaching improved over time and are better suited for adequate learning.

The survey was targeted at students at the Faculty of Aeronautics that were participating in online teaching in the winter semester of the academic year 2021/2022. The reason was to determine to which extent are these university students satisfied with their online classes, to see how many online platforms are being used in their study program and to give opportunity for them to express their ideas for improvement.

The questionnaire was administered online through Google Forms. It was distributed to the students through email addresses. The online questionnaire was composed of 18 questions, and it was estimated that it would take around 10 minutes to fill it out. The questionnaire features various types of questions such as multiple choice, matrix, and closed and open-ended questions. The questionnaire was composed of three parts, the first was dedicated to the gathering of demographical information, the second to the type of online teaching and the use of the most common platform, and the last section was addressing the satisfaction of students with the type of online teaching. At the end of the questionnaire, there was a space for their ideas and recommendations for improvement. The gathered data from the online questionnaires were evaluated by the means of descriptive statistics.

We received responses from 53 students during the period from 30 November to 2 December 2021. The survey was anonymous in order to keep the answers as sincere as possible. Of the total 53 students who participated in the survey, 17% of students were female and 83% were male. Most of the students were in their first year of study (55%), while the least represented was the fourth year of study with only 4%.

6 Results and Discussion

According to Fig. 3.1., we can see that 83% of students responded that videoconference is the most common method of online teaching, and it is used in more than 75% of classes. This is a considerable increase in comparison to the 2020 survey done by the Technical University of Košice, where the option that videoconferences are used in more than 75% of classes was chosen only by 7% of students. According to our study, the types of online teaching shifted from passive methods to active methods that are more interactive and feature interaction between the educator and the students.

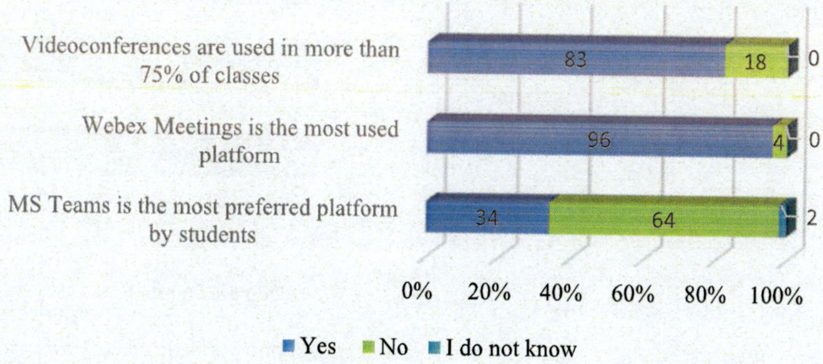

Fig. 3.1 Types of online teaching platforms

About the use of specific platforms, the results showed that the most used platform was Webex Meetings. The videoconferences conducted through a link to the personal room of the educator are used 88% of the time and videoconferences through spaces only in 9%. Videoconferences through personal room may save time for the educator, but at the same time, some students mentioned that this method is not very convenient for them as they have to search for the link in the emails. Only one student responded that Webex Training or MS Teams is the most common platform for videoconferences in their study program. Even though MS Teams was marked as the most used platform only by 2% of the students, it is the preferred platform for 34% of the students which showed a quite remarkable disproportion. Among the advantages of this platform, students mentioned that it operates more reliably, is multifunctional, is easy to navigate between the shared files and the ongoing videoconference, and provides notifications about the upcoming meeting and there is no need to search for the link for the meeting. The advantages of Webex Meetings through personal room were that it is easy to use and fast and it is the platform that is used for most of their classes. Features that were appreciated by students and are accessible on both platforms are the function of a whiteboard, access through browser and smartphone, and scheduling of meetings in the calendar. Some students mentioned that their preference was based on which platform they used as the first one; some students were first introduced to MS Teams and some to Webex Meetings either at their high school or university.

Figure 3.2 demonstrates inconsistency in the use of online teaching platforms by educators; it reveals that online education is performed on two and more platforms in 77% of cases, while only 23% of students have all their classes on one platform, which is Webex Meetings. Even though most of the students have to operate on more than one platform in attending their classes, the majority of them (42%) do not seem to mind this inconsistency, which may be caused by insufficient guidelines from the university.

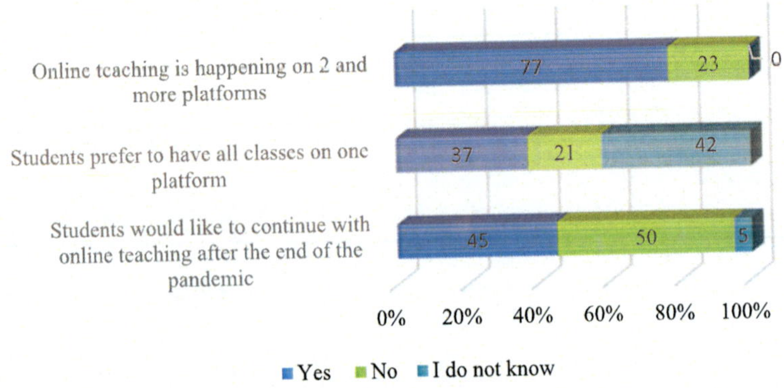

Fig. 3.2 Students' attitude to online teaching platforms

However, 37% would prefer to have classes on one platform and 21% of students do not want to have all their classes on just one platform. For students, it may be impractical to have classes on more than one platform, but on the other hand, this way each educator can choose their preferred method of teaching which they find the most convenient and with which they can work the best.

Considering the attitude of students toward keeping the online teaching methods employed even after the end of the pandemic was ambiguous, approximately half of the respondents (45%) would like to continue with some form of online learning and approximately half (50%) would not. Only 5% of students did not incline toward either possibility. According to these results, we can conclude that some aspects of online teaching could be preserved in the future for example in the case of external study. During this type of studying, students usually have a stable job, and they are studying along with going to work. In such cases, it could be very beneficial for these working students to have to opportunity to study online. In the open-ended question, some students in favor of face-to-face teaching stated that practical lessons cannot be sufficiently replaced by online teaching. However, with the current developments in digital technology, we may approach the level of sufficient online practical lessons that would comply with all students' needs.

In order to assess whether students at the Faculty of Aeronautics were more satisfied with online teaching in 2021 than students from all faculties in 2020, we have to compare Figs. 3.3 and 3.4. Figure 3.3 features the 2020 survey done by Technical University of Košice, and the Fig. 3.4 demonstrates the results from the current survey. Considering the first part of the figures on the left regarding the overall provision of information by the university is better, now 34% agree that they have enough information compared to only 18% in 2020. Less students marked the middle option of neither agree nor disagree meaning that they could have more time to evaluate all accessible options of gaining information.

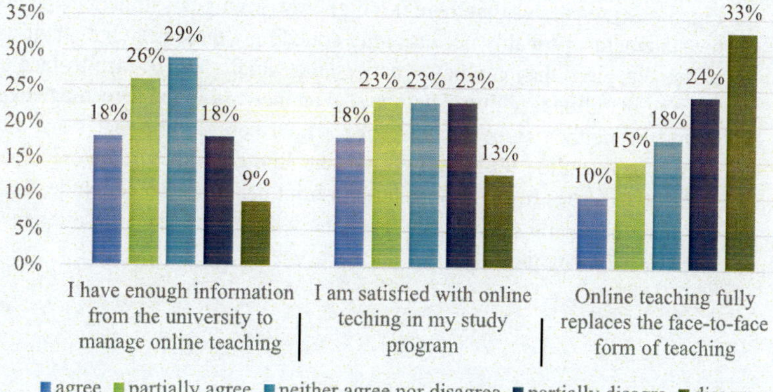

Fig. 3.3 Assessment of online teaching by students in 2020 at the Technical University of Košice

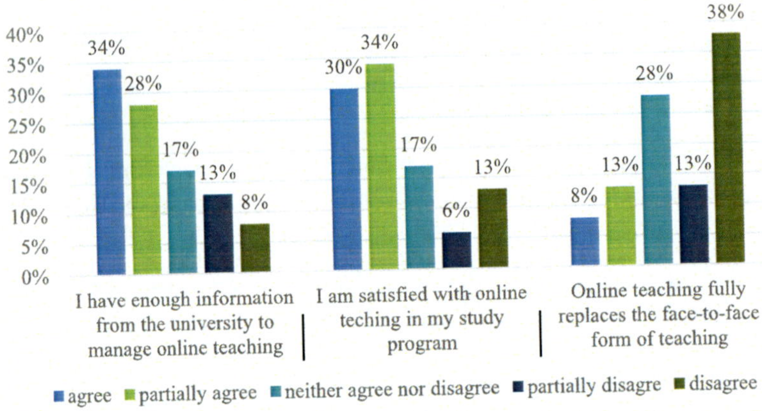

Fig. 3.4 Assessment of online teaching by students in 2021 at the Faculty of Aeronautics

Looking on the center part of the figures related to the satisfaction of the students with online teaching, the numbers are also better. In the present survey, 30% of students agree and 34% partially agree that they are satisfied with the way how online teaching is performed in their study program as opposed to the 2020 survey where only 18% of students agreed and 23% partially agreed with the statement. We can also observe a decline in the percentage of students who partially disagree with the statement while the percentage of dissatisfied students is kept intact. It is possible to attribute this improvement to the extended use of interactive online teaching methods such as videoconferences. Over time, the communication between students and educators could have also improved; however, students in the current survey mentioned that the communication should improve even more.

Lastly, the right part of the figures shows whether students think that face-to-face teaching is replaceable by online teaching; here, the answers are similar. However, Fig. 3.4 demonstrates that there is a higher percentage of students who disagree with the statements, probably because after spending almost year and a half using mostly online teaching, they can better understand which parts of face-to-face teaching are missing in online teaching. There is also an increase in the percentage expressing the option of neither agree nor disagree which can signify that some of the issues occurring in online teaching before were limited with further use or on the hand some other issues rose that forbid them from deciding. To conclude, the majority of the students would not want to continue with online teaching after the end of the COVID-19 pandemic.

7 Conclusion

Nowadays, digital technology is surrounding us everywhere even in the educational process. The modern developments in technology are about to change education as we knew it for decades; even now, the curricula have to adapt and copy the requirements for digital literacy that are posed by the job market. The balance between living life with and without digital technology, known as digital well-being, is a concept that is gaining importance. With the arrival of the pandemic COVID-19, the digital well-being of many people was threatened as the restrictions promoted social distancing causing many institutions including educational institutions to close their gates and switched to communication online.

The present study shows the views of students of the Faculty of Aeronautics on the quality of online teaching in their study program. For greater reliability, it would be beneficial to conclude such a study on a greater sample and possibly with the same students that gave their responses in the study from 2020. This study concludes that students at the Faculty of Aeronautics were more satisfied with online teaching in 2021 than students of the Technical University of Košice had been in 2020.

This could be promoted by the increased use of interactive methods of online teaching such as videoconferences, which are the most used in almost 90% of classes. The platforms used for such way of teaching were compared, and it seems that MS Teams is more adapted to the needs of online teaching, due to the assignment, test, and grading system it provides as opposed to Webex Meetings, which is more used at the Faculty of Aeronautics. Most students and educators prefer this platform; however, we would recommend paying more attention to the creation of spaces in Webex Meetings as it provides the benefits which can be found in MS Teams. The majority of the students do not agree that online teaching can fully replace face-to-face classes; however, some forms of online practices could be used further even after the end of the pandemic. Further studies could focus on progressive ways of maintaining the digital well-being of students and educators during the online teaching process, which will likely become one of the offered methods of studying at universities.

References

1. Kukreja, V. (2020). What factors impact online education? A factor analysis approach. *Journal of Engineering Education Transformations, 34*, 356–374. https://doi.org/10.16920/jeet/2021/v34i0/157180
2. Davis, F. D. (1989). Perceived usefulness, perceived ease of use, and user acceptance of information technology. *MIS Quarterly, 13*(3), 319–340. https://doi.org/10.2307/249008
3. Eksail, F., & Afari, E. (2020). Factors affecting trainee teachers' intention to use technology: A structural equation modeling approach. *Education and Information Technologies, 25*(4), 2681–2697. https://doi.org/10.1007/s10639-019-10086-2

4. Liao, Y. W., Huang, Y. M., Huang, S. H., Chen, H. C., & Wei, C. W. (2019). Exploring the switching intention of learners on social network-based learning platforms: A perspective of the push-pull-mooring model. *Eurasia Journal of Mathematics, Science and Technology Education, 15*(9). https://doi.org/10.29333/ejmste/108483
5. Chaka, J., & Govender, I. (2017). Students' perceptions and readiness towards mobile learning in colleges of education: A Nigerian perspective. *South African Journal of Education, 37*(1), 1–12. https://doi.org/10.15700/saje.v37n1a1282
6. Oh, E., Chang, Y., & Park, S. (2020). Design review of MOOCs: Application of E-learning design principles. *Journal of Computing in Higher Education, 32*(3), 455–475. https://doi.org/10.1007/s12528-019-09243-w
7. Khamparia, A., & Pandey, B. (2017). Impact of interactive multimedia in E-learning technologies.," rev. In *Enhancing academic research with knowledge management principles*. IGI Global. https://doi.org/10.4018/978-1-5225-2489-2.ch007
8. Pham, T. T. T., Le, H. A., & Do, D. T. (2021, 2021). The Factors Affecting Students' Online Learning Outcomes during the COVID-19 Pandemic: A Bayesian Exploratory Factor Analysis. *Education Research International*. https://doi.org/10.1155/2021/2669098
9. Floridi, L. (2015). The Onlife manifesto. In L. Floridi (Ed.), *The Onlife manifesto: Being human in a hyperconnected era* (pp. 22–30). Cham.
10. Büchi, M. (2021). Digital well-being theory and research. *New Media and Society*, 146144482110568. https://doi.org/10.1177/14614448211056851
11. Khoury, M., & Ioannidis, J. (2014). Big data meets public health. *Science, 346*(6213), 1054–1055. https://doi.org/10.1126/science.aaa2709
12. Twenge, J., Martin, G., & Campbell, W. (2018). Decreases in psychological well-being among American adolescents after 2012 and links to screen time during the rise of smartphone technology. *Emotion, 18*(6), 765–780. https://doi.org/10.1037/emo0000403
13. Frey, C., & Osborne, M. (2017). The future of employment: How susceptible are jobs to computerisation? *Technological Forecasting and Social Change, 114*, 254–280.
14. Burr, C., Taddeo, M., & Floridi, L. (2020). The ethics of digital well-being: A thematic review. *Science and Engineering Ethics, 26*(1), 2313–2343. https://doi.org/10.1007/s11948-020-00175-8
15. Aafaard, J. (2021). Beyond the rhetoric of tech addiction: Why we should be discussing tech habits instead (and how). *Phenomenology and the Cognitive Sciences, 20*, 229–572.
16. Zheng, Y., & Walsham, G. (2021). Inequality of what? An intersectional approach to digital inequality under Covid-19. *Information and Organization, 31*(1). https://doi.org/10.1016/j.infoandorg.2021.100341
17. Dennis, M. (2021). Digital well-being under pandemic conditions: Catalysing a theory of online flourishing. *Ethics and Information Technology, 23*(1), 435–445. https://doi.org/10.1007/s10676-021-09584-0
18. Centre for Humane Technology. (2020). *Digital well-being guidelines during the COVID-19 pandemic* [Online]. Available: https://humanetech.com/digital-wellbeing-covid. Accessed 2022/10/22.
19. Martinez, E. (2020). *Pandemic shakes up world's education systems*. Human Rights Watch [Online]. Available: https://www.hrw.org/news/2020/03/19/pandemic-shakes-worlds-education-systems. Accessed 2022/10/19.
20. Katz, V., Jordan, A., & Ognyanova, K. (2021). Digital inequality, faculty communication, and remote learning experiences during the COVID-19 pandemic: A survey of U.S. undergraduates. *PLoS One, 16*(2). https://doi.org/10.1371/journal.pone.0246641
21. Ministry of Education, Science, Research and Sport of Slovak Republic. (2020). Guidelines of the Minister of Education, Science, Research and Sports of the Slovak Republic regarding the interruption of classes in schools and school facilities (Usmernenie ministerky školstva, vedy, výskumu a športu SR vo veci prerušenia vyučovania v školách a školských zariadeniach). [Online]. Available: https://www.minedu.sk/usmernenie-ministerky-skolstva-vedy-vyskumu-a-sportu-sr-vo-veci-prerusenia-vyucovania-v-skolach-a-skolskych-zariadeniach/. Accessed 2022/10/20.

22. Student Council for Higher Education. (2020). *A guide to teaching during COVID-19* (Sprievodca výučbou počas COVID-19). [Online]. Available: https://srvs.eu/wp-content/uploads/2020/03/Sprievodca-v%C3%BDu%C4%8Dbou-po%C4%8Das-COVID-19.pdf. Accessed 2022/10/20.
23. European Commission. *Coronavirus: Online learning resources* [Online]. Available: https://ec.europa.eu/education/resources-and-tools/coronavirus-online-learning-resources_en. Accessed 2022/10/20.
24. Okabe-Miyamoto, K., & Lyubomirsky, S. (2021). Social connection and well-being during COVID-19. In *World happiness report 2021* (pp. 131–152). Sustainable Development Solutions Network.
25. United Nations. (2020). *Education during COVID-19 and beyond*. United Nations.
26. Čopková, R. (2020). *Distance teaching method at Technical University of Košice (Dištančná metóda výučby na TUKE)*. Technical University of Kosice, Department of Engineering Education.
27. Institute of Computing Technology TUKE. (2020). *Webex* [Online]. Available: https://uvt.tuke.sk/wps/portal/uv/sluzby/webex. Accessed 2022/10/21.
28. Webex Help Center. (2021). *List of features for Webex meetings 40.11 to 41.2* [Online]. Available: https://help.webex.com/en-US/article/n15ck46/List-of-Features-for-Webex-Meetings-40.11-to-41.2. Accessed 2022/10/21.
29. Webex Help Centre. (2020). *Get started with Cisco Webex training* [Online]. Available: https://help.webex.com/en-US/article/xqsxxt/Get-Started-with-Cisco-Webex-Training. Accessed 2022/10/21.
30. Webex Help Center. (2020). *Test and grade in Cisco Webex training* [Online]. Available: https://help.webex.com/en-US/article/n82bh9v/Test-and-Grade-in-Cisco-Webex-Training. Accessed 2022/10/21.
31. Microsoft. *Microsoft Teams learn and grow together* [Online]. Available: https://www.microsoft.com/en-us/microsoft-teams/education. Accessed 2022/10/21.

Chapter 4
Digitization, Personalization, and Automation of Business Processes

Romana Hricová

1 Introduction

Today's fast time brings with it too much information, which companies operating on the market must know how to use correctly. Every enterpreneur who wants to maintain or improve his market position and be successful must know how to process the available information and how to use it properly. Communication and its methods between entrepreneurs and their customers, whether existing or potential, have also experienced major changes. The old ways are not only unattractive but also especially financially demanding and do not bring the desired effect. Therefore, companies must digitize the available data about their customers, as this brings benefits not only to the companies themselves but also to the customers. A customer who receives a "made-to-measure" offer is included in the category of satisfied and especially loyal customers.

Nowadays, many businesses are running digitization initiatives or are planning to do so [1]. Digitization was identified as one of the main trends, which change business and society as well [2]. Digitization is also important for choosing the right marketing, which is one of the most important parts of business management. Many companies and their managers do not realize that without marketing, their company will not function as they imagine and, above all, will not use all the possibilities that are available. They do not take into account that marketing is a relatively fast-developing industry, which is subject not only to fashion but also to the changing tastes of customers and, above all, can use the modern achievements that today's times bring to the benefit of the company. Therefore, marketing has

R. Hricová (✉)
Technical University of Košice, Faculty of Manufacturing Technologies with seat in Prešov, Bayerova 1, 080 01 Prešov, Slovak Republic
e-mail: romana.hricova@tuke.sk

L. Knapcikova, S. Zohrehvandi (eds.), *9th International Conference on Mobility, IoT and Smart Cities*, EAI/Springer Innovations in Communication and Computing,
https://doi.org/10.1007/978-3-031-28225-6_4

become one of the main representatives and tools of communication in advertising. The development of technology is moving forward in various areas, so marketing communication does not lag behind. Day by day, it increases the connection between people, businesses, and institutions through modern information tools. The time when marketing was focused primarily on the sale of products and services has, to a certain extent, disappeared. The old way of promoting goods has become unattractive for the company [3].

In order for businesses to be able to respond correctly to the needs and demands of their customers, it is necessary for entrepreneurs to realize that today's business is different than it was only in the recent past. Especially in recent years, when many businesses were forced to move to the online space in order to survive at all, the competition has become so close that sometimes a few clicks on the computer are enough for the customer to change the store where he decides to shop. It is literally a matter of minutes, if not seconds, before a business loses a potential customer to a competitor. In addition, the differences between the products and services themselves are evened out. Therefore, a dissatisfied customer does not hesitate for a long time and finds another seller.

Businesses have counted on advertising for a long time and have largely equated advertising campaigns with marketing. However, that is slowly but surely ceasing to be true. Today, the average person is exposed to a lot of advertisements on a daily basis. Their number depends on whether he watches TV, reads newspapers, listens to the radio, surfs the Internet, uses a smart phone, or walks around town. Studies show that the number of advertising impressions ranges from an average of 250 per day to an upper limit that ends at 3000 [4]. Therefore, customers who are overwhelmed in this way become immune to advertising, and businesses have to find other ways to reach to receive to the customer.

Knowing how to reach the right customers at the right time and especially with the right offer has become a challenge for companies on the market. More and more demanding customers who have access much faster and, above all, to a larger amount of information than was the case not long ago choose and compare, and as it turns out, price is not always the decisive factor. Parameters such as the customer's overall impression of the company's approach, how communication took place, how quickly feedback was provided if necessary, and even such a factor as the perception of the brand as such gradually came to the fore in relation to the product, which is always paramount.

In addition, the events associated with the global COVID-19 pandemic, when people were forced to work from home and also to shop from home, significantly affected mainly businesses selling consumer goods, which had to – if they wanted to stay on the market – move to the online space. All this has created an exceptionally suitable environment for the rapid growth of digital marketing. This is how digital marketing (and the marketing automation that goes hand in hand with it) becomes a very important tool for more effective, faster, and, above all, more targeted customer outreach.

2 Problem Definition

Today's businesses are trying to take advantage of the available options, and the result is that they have a lot of information in different places. Part of the data is in databases, part in paper form, part on social networks, and part only in the heads of workers. All these not only make the search for the necessary information difficult but also mainly slow down the company when communicating with the customer.

Digital marketing has an impact on improving business performance, but there is also a negative impact of adopting digital marketing on improving marketing performance. It is important to realize that before investing money in digital marketing tools and their application, enterprise management should create a business culture emphasizing the strategic importance of digital marketing, carefully plan digital marketing activities, and develop an adequate digital marketing analysis system. This will enable him to continuously optimize his digital marketing activities and increase his effectiveness [5].

Customer loyalty is often considered one of the decisive factors in determining the fate of organizations. In the digital age, organizations should think about advanced strategies to increase their competitiveness and market share by harnessing the potential of digital content and improving their digital capabilities. Recently, there have been many digital media available to organizations to build their brand, reach and retain their target customers, and promote their products. Digital media represented a real revolution that made it difficult for organizations to survive in the business environment without investing in these means [6].

The presented contribution is aimed at showing the importance of digitization, subsequent personalization and automation in the company, how the aforementioned activities can increase the effectiveness of not only marketing communication, and what benefits it will have not only for the company itself but also for its existing and potential customers.

3 The Consumer and the Digital Market

Consumer behavior in the market is changing. Already today, the share of consumers in the EU who buy goods or services via the Internet is more than 60%. It is stated that, e.g., since 2004, when it was 21% in 10 years, this share increased almost 2.5 times to 50% in 2014 [7]. From 2017 to 2018, the growing trend of e-commerce use continued, which was recorded in the last decade. Approximately 60% of consumers made online purchases [8]. It is also important that the proportion of individuals who last bought online in the past 3 months is gradually increasing and the share of those who last bought online in the past 3–12 months, which means that the frequency of e-commerce transactions is also increasing.

In 2017, 19.5% of all companies (with 10 or more employees) provided online sales, and e-commerce accounted for 17.4% of the total turnover of companies

[8]. These numbers have been relatively stable for several years. However, significant differences remain between EU countries: while in Ireland, Sweden, and Denmark, 3 out of 10 companies provide online sales; in Romania and Bulgaria, it is only 1 out of 10. The distribution of the same indicators by sector shows that the share of companies providing online sales, as well as the share of online turnover, is the largest in the area of accommodation services (68.1% and 31.6%), followed by retail (28.0% and 10. 7%), while construction is the least involved in e-commerce of all industries (the share of companies is 4.3% and the share of online turnover is 1.8%).

3.1 The Impact of the Pandemic on Digitization

The COVID-19 pandemic has forced people to change their shopping habits, so many more people have moved into the online space. E-shops, under the pressure of new orders and an increase in the number of customers, had to adjust their processes so that they were fast, efficient, and user-friendly, since the fierce Internet competition goes beyond geographical borders. The data show that some Slovaks continue to shop online even after the easing of measures, as payment volumes in online stores far exceed pre-crisis levels or levels reached in the pandemic year 2020.

E-commerce already experienced a strong influx of customers in 2020 due to the pandemic, and as data for payments show, as a result of significant restrictions on brick-and-mortar stores, the volume of payments in e-shops increased by 40% year on year, especially in Slovak ones (+80%) since the structure of purchases has changed due to the pandemic [9].

Even in 2021, purchases on the Internet remained high. They were even higher than in 2020, since at the beginning of 2021, measures limiting mobility and contact industries, including retail, were again adopted in Slovakia. The volume of purchases peaked in March 2021, when it was up to three-quarters higher than in the same period of 2020 and even more than twice as high as in March 2019. After the subsequent improvement of the pandemic situation and the related easing of measures, the volumes are continuously decreasing, but they are still above the levels of 2020.

It was the pandemic and the associated measures preventing personal contact that forced many Slovaks to try purchasing goods via the Internet for the first time, and some of them continue to use these services even after the easing of measures. Significant growth can be seen, for example, in food deliveries, where in some weeks the volume of purchases before the pandemic has been exceeded several times. In addition, purchases in the category of clothing and footwear or household goods are also higher.

Slovaks buy almost everything on the Internet, yet the online market recorded a year-on-year decrease of 28% in the first quarter of this year (2022) [9]. The reason is the extremely high increases in Slovak e-commerce during the pandemic, i.e., in the previous period, and the return of customers to brick-and-mortar stores. At the same time, in connection with the current situation, customers began to think more about what they buy.

Uncertainty and worries about the future have contributed to the fact that Slovaks shop less on the Internet. Consumers are becoming more cautious when making purchases, and the question remains how long they will continue to do so. Sellers who have boldly invested in further development in anticipation of further e-commerce growth may be threatened by this behavior. In general, however, significant changes during the last turbulent years are excellent news for e-shops.

As for the types of goods purchased, electronics have always been the most important category that consumers buy on the Internet. However, compared to the past, the differences in the shares of individual categories have started to decrease. Electronics currently make up approximately 30% of all online purchases, while 3 years ago, it was up to 60%. The last 2 years and especially the frequent lockdowns have taught customers to buy almost everything through e-shops.

The pandemic literally shot some segments to astronomical growth. For example, a significant increase was recorded in categories such as auto-moto, cosmetics and health, sporting goods, and home and garden. The best example is greenhouse, which is one of the fastest growing product categories in online shopping. A few years ago, few people would have imagined buying a greenhouse from an e-shop.

Even before the pandemic, experts warned that the online market will gradually cease to be the domain of only certain categories and consumers will gradually start buying literally everything through e-shops. The pandemic has extremely accelerated this trend. Some experts claim that we even jumped 7 years ahead during it. At the same time, it is a unique chance for e-shops to expand their assortment.

The Association of Intelligent Industry – Industry4UM conducted the fourth annual survey on the level of digital transformation in industrial enterprises at the turn of August and September 2020. Compared to 2019, the survey revealed stagnation in the digitization of industrial enterprises. The results of the survey were also influenced by the unprecedented situation associated with the COVID-19 pandemic. As shown in Fig. 4.1, up to two-thirds of companies (out of 57 contacted companies) realize that they would be able to handle the impact of the corona crisis better if they were digitized [10].

As soon as companies realize the potential of e-shopping and start digitizing processes in the company, it is high time to start working on marketing automation as well.

4 Automation of Marketing

Although managers and marketers always appreciate the endless flow of fresh information about consumers, which they then use to improve their products and services, continuous process improvement in marketing automation is essential [11].

Of course, marketing automation is not a tool that will immediately bring an increase in sales, but it should be a carefully thought-out part of a comprehensive

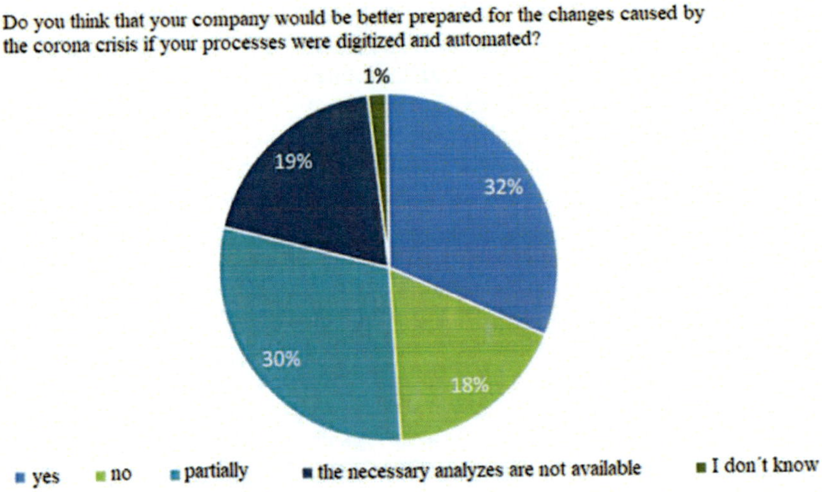

Fig. 4.1 The result of the survey. (Source: https://industry4um.sk, published: 16.11.2020 [10])

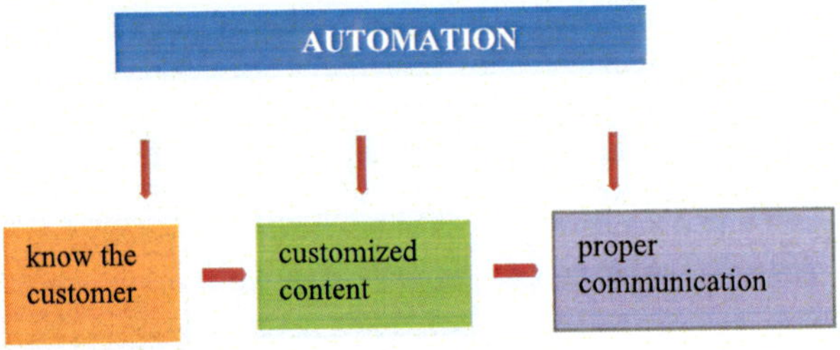

Fig. 4.2 The process of automation

marketing strategy. Businesses need to realize that the general fact remains that it is easier to keep a customer than to find a new one. In addition, a satisfied customer will actually advertise the company for free, because he is happy to share his good experience, e.g., with friends. So it remains for the company to build a relationship with the customer based on respect and loyalty.

Figure 4.2 shows how automation works. A business that wants to automate its marketing must first of all know its customer perfectly. This means that there must be a folder where the basic customer data is. In addition to name, address, and other identification data such as IČO (identification number), DIČ (tax identification number), and account number, the company should also know such data as who is authorized to conclude contracts on behalf of the company, who decides on purchases related to the company, or who and how can influence them. He must also know all the available information that is available in the company about the customer and centralize it so that it is accessible in one place, and there is no need to search for it in several places (e.g., from social networks, company databases, employee notes, etc.).

In the second step, the message content is created tailored for that specific customer. If the company has many customers, it is possible to divide them into groups that have the same characteristics and can be addressed in the same way. Alternatively, make only minor adjustments. In case this is not possible, it is necessary to really create a unique message for each customer, which will be more or less different from the others.

The last, third step is about proper communication. It is necessary to know the customer well enough for the company or marketer to find the right communication. It means that the customer must be contacted so often that he does not forget about the company, but not so often that he finds it annoying and thus does not discourage him.

In this direction, also on the side of merchants, e.g., retailers most often encounter competitors who pressure consumers through constant commercial calls or messages, while over a third have encountered competitors who falsely advertise a product as being available for a limited time only or write misleading reviews that constitute disguised advertising or veiled attacks on competitors. Customers are often offered products that are supposed to be free, when in fact there are significant fees associated with them, and often the customer is also sent an unsolicited product and then asked to pay for it. All of these reasons play an important role in automation, as the business must proceed in such a way as to acquire the client and not deter him.

Many marketers still have significant concerns about automation – especially given the broader industry push to incorporate more human connection and personalization to improve response. No one wants to receive generic emails that are addressed to a large number of clients.

In 2019, a survey was conducted on a sample of 300 respondents regarding the use of automation [12]. The results of this survey showed several interesting facts, but the most important was the answer to the question whether the company uses automation tools. Up to 75% of respondents answered positively. It means that companies already accept automation; the question remains what companies consider to be marketing processes that can be successfully automated.

If we stop at email marketing automation, then it is important and necessary to segment customers as mentioned above. If segmentation is not used or used in an inappropriate way, the position of the company in the eyes of the customer can deteriorate and the overall image of the company can be damaged. So the importance lies precisely in the effective segmentation of customers. For example, if a customer has just purchased a new personal computer, it is very likely that they will not be in the market for personal computers for some time. Sending him an offer of new computers would not only be pointless, but counterproductive and certainly not look professional to the customer. On the other hand, offers for additional goods such as mouse, headphones, speakers, memory cards, printer, special keyboard, or even a gaming chair can be useful for him. And such activities can put the company in the light of a caring and professional businessman who cares about his customers and has an overview of them. Therefore, additional filtering of email contacts into specific groups can have a big impact in this regard and ensure that automated processes generate the optimal response.

Automation brings with it many benefits, which mainly include the following:

- Time saving
- Increase of revenue
- Customer retention
- Monitoring and tracking of marketing campaigns
- Shortened sales cycle
- Direct interaction with the customer
- The possibility to get quick and relevant feedback
- Improved targeting of messages
- Improved customer experience
- Getting more leads
- Improved efficiency in marketing
- Closer cooperation between marketing and sales
- Efficiency of cost
- Ability to combine data from different channels
- Better profit in total

5 Personalization

Once a business knows what to expect from automation, personalization follows. Personalized marketing thus becomes a necessity for marketers who want customers to return and make repeat purchases. The reasons for personalization are as follows:

- Up to 81% of consumers want businesses to get to know them better.
- 75% of consumers expect personalized communication from businesses.
- 78% of Internet users said that relevant personalized content increased their interest in buying.
- 65% of consumers said they are likely to switch brands unless a business personalizes communications with them.
- More than 50% of consumers are willing to hand over personal data if the company would use it for responsible personalization [13].

In the first step, which is personalization, it is necessary for the marketer to answer the following questions:

- Do I do personalized email marketing? If so, how much time will I spend preparing it?
- How many website visitors will I convert into real customers?
- Can I identify how many customers leave the cart without completing the order?
- Do I somehow work with customers who abandon the cart without ordering?
- Do I track OR (Open rate) and CTR (click to rate) from the newsletter or mailing?

- Would personalization improve OR and CTR values?
- Do I reactivate those customers who leave the site without any action?
- Do I do personalized upsell and cross-sell?

It is important to know that all these actions can be automated.

Ultimately, the job of personalization is to find out what the customer wants. The company and its employees should abandon the idea that they themselves know best what the customer wants, needs, and demands. This is a very common problem that marketers rely on their previous experiences (either general or their own) and believe that they know better what the customer wants than the customer himself. However, this can be a very wrong idea.

The second option is for the business to ask its customers and potential customers what they want. Whether in the form of surveys, questionnaires... Even if the company obtains more accurate information that it can use further, it still may not be completely accurate. A poorly formulated questionnaire, an inappropriately selected survey sample, a respondent who does not answer honestly, all of this can distort the results, and the company will thus build on something that is not entirely true. This situation is very clearly portrayed by, e.g., surveys regarding the closing of shops on Sundays. Various surveys indicate that the majority of Slovaks are in favor of closing retail stores on Sundays. Some surveys report up to 80% [14]. However, just visit any supermarket or shopping center on a Sunday and you can immediately see that they are full of shoppers. This suggests that respondents largely answered as they thought was expected of them and was humanly correct. However, if questions were asked differently, e.g., whether the respondent is in favor of the opportunity for salespeople to earn extra while working on Sundays, many would answer yes, which would ultimately mean that they are actually in favor of opening stores on Sundays as well.

For this reason, the third and best option for personalization is that the marketer should observe how the customer actually behaves and draw conclusions from it.

6 Automation and CRM

With automation, it is advisable for businesses to operate in a customer relationship management (CRM) system. CRM is a way to approach the customer. It contains various methods of customer care and will allow you to take customer relations to a new level. Among his main and primary tasks is to create long-term and good relations with the customer.

In today's technological age, CRM is aided by the fact that various programs are available, thanks to which the company can increase the productivity and efficiency of work processes. It works on the basis of automation, centralization, and last but not least synchronization of entered or inserted data CRM software is now commonly used to organize, automate, and synchronize sales, marketing, and customer service. This means that the business can thus centralize all available information about its customers and also about the ongoing interaction between the business and the customer. So, in addition to contact data, he also has available communication that took place either in the form of emails or phone calls, transactions made, business offers, and possibly other important information and documents. Subsequently, CRM software will facilitate access to this information, and because modern CRM systems work online or in the cloud, this information will be real-time and always up-to-date. This will make it possible to work with the customer better and, above all, more efficiently.

6.1 Content of the CRM Software

It cannot be assumed that there is one software that fits all businesses. On the contrary, CRM software will differ depending on the industry in which the company operates, how big business is, how many customers it serves, and last but not least whether it has several areas of operation or just one.

In general, it can be stated that CRM software should have the following modules:

- Central directory of contacts
- Tool for recording communication with clients
- Tool for recording business opportunities
- Tool for recording projects and orders
- A tool for planning activities and tasks [15]

The possibilities of CRM software are very large, and it depends on the specific company what it decides for. To begin with, it is suitable if companies choose only a part, which they can gradually expand after appropriate application. It is best to start with emails and sms marketing, which is direct marketing, the purpose of which is to inform about news, improve relations with the customer, build their trust, and arouse interest in the company's products or services. The task is, on the one hand, to support the repeat order of already existing customers and, on the other hand, to establish a relationship with new customers. It should take place in the following steps:

Step 1 – Creation of dynamic personalized content

The software should have options to set the sending of emails automatically and also to schedule them. It is good if the company develops several variants and then tests them. Individual variants may differ in content, template, message subject... In this step, the marketer should definitely not forget to assess the technical parameters of the services. An email service provider (ESP) is a service that allows marketers to send email marketing campaigns to a list of users (subscribers). Subscribers join these lists by subscribing to receive marketing messages. Subsequently, and also suitable, the marketer starts with a request for analysis of the list of recipients and finds out whether the ESP performs this analysis automatically for all customers and lists of recipients or only on request. The analysis serves not only ESP but also the customer to reveal nonexistent addressees, suspicious addresses, generic addresses, or temporary addresses and their proportions in the list.

In addition, the analysis of recipients will help reduce the volume of sent messages and thus save money, but above all it will also point out potential problems in the process of email data collection. ESP, on the other hand, obtains an overview not only of the up-to-date data but also of the customer's processes and should be able to assess the reputation and prepare a strategy to ensure deliverability based on this analysis. If ESP considers the list risky, it should request additional information from the customer and refuse to provide services if the criteria are not met.

Step 2 – Analysis of the campaign's spam score

For the analysis, it is possible to use one of the previous campaigns or a campaign that the marketer is going to send for testing via the service. It should not test the sensitivity of the system and its limits, but only the content.

Step 3 – Test mailing to a test list of recipients

The test list should reflect the distribution of the entire recipient database among the individual email services and should be at least hundreds of addresses, but preferably thousands. The marketer should create several custom mailboxes for each email service (e.g., Gmail, Yahoo, pobox.sk, azet.sk, zoznam.sk, etc.) according to importance. It adds these addresses to the test list and has a ready basis for the deliverability test. This is followed by the preparation of the campaign in the usual form and its distribution at the usual time – it is necessary to realize that the vast majority of recipients in the test list are real customers.

Step 4 – Verification of results

After 24 hours, it is advisable to check the results, that is, to which mailboxes the campaign was delivered, where it was included in the inbox, where it was included in bulk mail, where it fell into spam, and where it did not arrive at all.

There are delivery monitoring services that can save both work and evaluation, but they may not support Slovakian free mail services. Self-measurement takes more time, but the company receives more detailed information. In addition, the values measured by ESP itself can be significantly distorted and may not reflect real deliverability.

Step 5 – Obtaining information from the distribution of test reports

The test sending of emails should also provide basic information about ESP's ability to evaluate nondelivery of messages, the so-called soft and hard bounce, i.e., temporarily unavailable and permanently unavailable addressees. Even if an analysis of the list was performed at the beginning, it is likely that some addresses could not be verified completely, or only with a certain probability, or addresses may be unavailable again, e.g., due to full mailboxes.

A good system will give accurate information about the type of message undeliverability and can distinguish between temporary and permanent undeliverability. Recipients who are temporarily undeliverable must be eliminated in case of repeated nondelivery, and the system should automatically ensure this.

Failure to follow this rule, which varies by email service, has a significant impact on reputation and deliverability.

Some systems will display detailed server responses with the reason for undeliverability. Here, you can also find possible warning signals noticing about the blocking of the ESP or the sending address. If the system does not provide detailed error answers, it is a signal that it is a technical solution that has fundamental limitations or that it wants to keep this data secret.

Step 6 – Feedback from recipients
From the campaign statistics, the marketer should also learn such information as, for example, which customer marked the mail as spam. ESP receives this information from email services using feedback loop (FBL). Not all email services provide FBL, but the largest ones such as Outlook.com (Hotmail), Yahoo, AOL, or Zoznam operate a publicly available FBL that allows ESP to process information about spam reports. Information about the number of spam reports has a significant impact on the delivery of future campaigns. It is essential that recipients who mark an email as spam are not contacted by further campaigns in the future.

7 Conclusion

Reaching the right customers at the right time is the desire of many companies. No one wants, and especially in today's challenging times, to unnecessarily invest in marketing expenses whose return is not guaranteed or is too low. Therefore, companies and their marketers are looking for ways to target their customers and potential customers with the greatest possible return. Thanks to digitalization, companies are opening up possibilities that were not technically possible just a few years ago.

Digitization makes it possible to put all information about customers in one place, where they can then be personalized and create specific content for a specific customer. Targeting has many benefits, from making the customer feel special to controlling marketing spend, where the business knows exactly how much it is investing and what the response is. That is, whether the response is adequate for the costs incurred. If necessary, it is also possible to react quickly and make changes. Corrections are even desirable, because the business environment changes rapidly, businesses are created and disappear, and clients come and go.

When a business is aware of all these facts, it can automate its marketing activities so that the marketer can devote his attention and time to other activities. A business that has automated marketing solutions can track, document, and, above all, analyze in relative detail those who have visited the company's website and those who have registered to receive news. The better the company knows its customers and potential customers, the better it has an idea of their needs and can send them targeted advertising and thus increase sales success. And higher profit goes hand in hand with that.

Since the effort to digitize is pan-European and the European Commission plans to make Europe a digitally sovereign continent by 2030, it is time for businesses to not only deal more and more actively with digitization, personalization, and automation but also take into account the various risks that this results, for example, in the field of cyber security. The more digitization increases, the greater will be the danger arising from cyber security. In addition, not only the requirements for increasing digital skills in the future will result from the needs of businesses, but also people themselves will be faced with the necessity to increase and expand their digital skills. It is therefore necessary to continue to focus on this area, to look for opportunities to expand digitization both in businesses and among residents, and to create safe, efficient, and, above all, sustainable options.

Acknowledgments This publication was supported by the VEGA 1/0268/22 granted by the Ministry of Education, Science, Research and Sport of the Slovak Republic.

References

1. Heberle, A., Lowe, W., Gustafsson, A., & Vorrei, O. (2017). Digitalization canvas – Towards identifying digitalization use cases and projects. *Journal of Universal Computer Science, 23*(11), 1070–1097. ISSN 0948-695X.
2. Parviainen, P., Tihinen, M., Kaariainen, J., & Teppola, S. (2017). Tackling the digitalization challenge: How to benefit from digitalization in practice. *IJISPM – International Journal of Information Systems and Project Management, 5*(1), 63–77. https://doi.org/10.12821/ijispm050104
3. Behúnová, A., Behún, M., & Knapčíková, L. (2022). Using industry 4.0 tools in the tmplementation of modern event marketing. *Mobile Networks and Applications*. https://doi.org/10.1007/s11036-022-02047-w
4. Poliakova, M.: Vědci spočítali, kolik na nás během jediného dne působí reklam. https://radiozurnal.rozhlas.cz/vedci-spocitali-kolik-na-nas-behem-jedineho-dne-pusobi-reklam-6234473
5. Mahmutovic, K.: The impact of digital marketing orientation on hotel marketing performance. *Interdisciplinary management research XVII (IMR 2021)*, pp. 1211–1232. ISSN 1847-0408.
6. Mohammad, A. (2022). The impact of digital marketing success on customer loyalty. *Marketing and Management of Innovations*, (3), 103–113. https://doi.org/10.21272/mmi.2022.3-09
7. https://ec.europa.eu/info/sites/default/files/ccs2015scoreboard_sk.pdf. Accessed 2022/10/20.
8. EÚ Európska komisia, Generálne riaditeľstvo pre spravodlivosť a spotrebiteľov, Hodnotiaca tabuľka spotrebiteľských podmienok: spotrebitelia na jednotnom trhu ako doma: rocnik 2019, Publications Office, 2019. https://data.europa.eu/doi/10.2838/745984
9. Slovenská sporiteľna Homepage. Zdroj. https://www.slsp.sk/sk/aktuality/2021/10/25/e-shopom-sa-dari-slovaci-si-na-online-nakupy-zvykli. Accessed 2022/10/21.
10. Industry 4UM Homepage. https://industry4um.sk. Accessed 2022/10/21.
11. Jeknic, J., & Kraut, B. (2015). Cloud services and marketing. *8th international convention on information and communication technology, electronics and microelectronics (MIPRO)*, pp. 1492–1498.

12. Hutchinson, A. *SMT's state of marketing automation survey 2019 – Part 1: Current state of automation*. https://www.socialmediatoday.com/news/smts-state-of-marketing-automation-survey-2019-part-1-current-state-of/551755/
13. Talarovič, R. Nastavte svoj biznis na autopilota. Odborný on-line seminár, 9.február 2021. https://www.npc.sk/sk/events/nastavte-svoj-biznis-na-autopilota-3/
14. https://www.aktuality.sk/clanok/792087/prieskum-az-80-slovakov-chce-mat-obchody-v-nedelu-zatvorene/.
15. *CRM system*. http://www.crm.sk/crm-software/

Chapter 5
Smart Solutions for an Improved Experience in the Process of Buying Slovak Products

Ján Ganobčík, Katarína Gubíniová, Martina Jantová, and Gabriela Pajtinková Bartáková

1 Introduction

The current trends in customer care system represent a continuous shift from customer relationship management to customer experience management. This shift is caused because of the fact that offering only single product and/or service for the customer in the current competition situation on the B2C market is not enough. It follows that organizations should focus their attention on the quality of consumer experience and the various experiences that they have with the organization. It has never been more difficult to achieve and maintain organizational success in the business activity of the current organization only through differentiation in the most important instruments of the marketing mix (from this point of view), i.e., product and price. The brand of the product has to imply not only the quality of the product and its features but also the quality of different customer interactions with the organization (by which we can include speed, simplicity, effectiveness, efficiency, and convenience of all processes in the organization evident/covert for the customer), which can either create or break the relationship with organization. The originality of the paper stems from combining the theoretical knowledge about consumer care system combined with current trend in many managerial areas – smart solutions, in the paper represented in the process of buying Slovak products.

J. Ganobčík · K. Gubíniová (✉) · M. Jantová · G. Pajtinková Bartáková
Comenius University Bratislava, Faculty of Management, Department of Marketing, Bratislava, Slovakia
e-mail: jan.ganobcik@fm.uniba.sk; katarina.gubiniova@fm.uniba.sk; martina.jantova@fm.uniba.sk; gabriela.bartakova@fm.uniba.sk

L. Knapcikova, S. Zohrehvandi (eds.), *9th International Conference on Mobility, IoT and Smart Cities*, EAI/Springer Innovations in Communication and Computing,
https://doi.org/10.1007/978-3-031-28225-6_5

1.1 Customer Experience and Artificial Intelligence

During the last 10 years, the customer experience has received remarkable attention in both marketing academics and managerial practice. Both academics and managers responsible for the implementing the marketing strategy believe that customer experience is key to the organization's competitiveness among the direct/indirect competitors [25], and academics refer to it as the core of modern/current marketing management [11, 23].

Customer experience is the altogether experience a (final) customer has with any organization through their interchanges and expectations about the brand [32, 46]. Academics studies identify four components of the customer experience: (a) mental, (b) emotional, (c) physical and sensory, and (d) social components [20].

In accord with Keiningham et al. [18], mental components of the customer experience belong to the functionality, speed, and accessibility of the product. Additionally, earlier academic (and also empirical) studies have stressed the emotional components of customer service, which tend to be involved in nature [20]. The feelings of the customer can be positive or negative, such as joy, regret, anger, outrage, joy, or surprise [18].

In comparison, the physical and sensory components of the customer experience are frequently distinguished between elements in offline and online environment. Offline encounters include traits such as artifacts, lighting, layout, and signage [21], while online experiences include technology-related traits such as a user-friendly interface and obvious design [18]. Finally, the social components of the customer experience refer to the influence of other people such as family members and relatives, friends, and the customer's wider social networks [46].

Social elements also involve the social individuality of the customers or the mental identity of how they themselves look at a particular product or service [18]. These elements of the customer experience also influence the perception and acceptance rate of artificial intelligence (AI). According the Gartner's study, AI, such as machine learning, natural language understanding, and natural language processing, can help to analyze customer sentiment and customer feedback at a scale, accuracy, and speed that humans cannot achieve [8]. This suggests that AI has the enormous potential to become one of the main tools for organizations to continuously improve the customer experience on the B2B market and thus remain enough competitive [29].

In retail, AI technology is often used in combination with other technologies such as augmented reality, robotics, computer vision-based image recognition, predictive inventory, and smart inventory, and in the customer service processes [40]. For these technologies to successfully improve the customer experience, there exists a need to have a thorough comprehending of the customers, containing their choices and expired (positive) experiences. Any AI application can support expediting this understanding, because AI tools use historical customer data and created profiles to learn, respectively, to improve how to communicate with customers [33]. However, for proper use of any AI tool, it is important to emphasize quality.

The quality of such services is commonly defined as the distinction between anticipated and received level of the service quality and is evaluated according to how customers perceive an organization's portfolio offering [34]. This concept of service quality has its origin in the theory of disconfirmation of expectations [5], where the assessment of service quality is the outcome of a comparison between the perception of the received service and the previous anticipations of what the service provider should provide [16]. The existing collection of research in the field is rich in studies of interpersonal service quality [38, 43], while there is a lack of academic papers on customer reactions to automatic services, specifically AI-enabled services [37]. As AI-enabled services are typically built on self-service technologies, service quality in the context of AI-enabled services is likely to be consequently different from interpersonal services.

The quality of AI-enabled services depends largely on the level, and quality of personal information an organization is able to gather about its (existing/perspective) customers. Although most of these data are usually not sensitive, the mixture of seemingly nonsensitive personal information (such as marketing activities and preferences) can lead to a large user profile that, with unsatisfactory protection measures, would allow fraudsters to create false identities [15]. For these reasons, the security and reliability of services are especially important.

According to Saratchandran [41], assuming the ability to have "unbiased" interactions with customers through smart solutions, the trustworthiness of customer services increases. While AI-enabled services are much more likely to replace current biases with new ones, they are much more scalable than traditional services and have the potential to help large numbers of customers together. Chatbots and other customer service instruments are growingly being used as an automated and possibly effective direction to improve the customer experience [44].

As many AI-enabled services has their foundations a self-service model, a thoughtfully designed user interface is often defined as a crucial success reason for such services [17]. Former academic works confirm that the technical and functional quality of services influence the way customers assess organizations, respectively, brands [6]. In the case of lack of other information, the type of technology and how it is implemented by the service provider perform act as a substitution for its quality from the consumer's perspective and support consumers establish an initial level of confidence.

1.2 Technological Innovations in Retail

In recent more than 10 years, technology has involved customers in the process of daily use of social networks [10], mobile applications [26], and chatbots [14]. Consequently, technological consumer engagement helps to develop different social relationships among automated services and consumers [45]. Within the broader (eco)system, technical innovation can be considered an "actor" in the process of engagement and at the same time a "contributor" to co-creative activities, especially given the capabilities of machine learning. With this in mind, the customer service experience within the future has the tendency to be shaped by the level to which technology involves to a various social degree. It is imperative to examine the status of its social interactions within technological innovation in order to define the role that such system can operate in customer stimulation. Most of today's self-service technologies have a deficit of the ability to involve consumers socially. Therefore, technologies that are truly competent of involving in personalized social connections and developing various relationships with (final) consumers at the cognitive, affective, and behavioral levels are likely to have associations for the consumer in terms of customer experience. New technologies can provide a socially engaging experience similar to personal. However, in this direction, the necessary phases need to be done in order to comprehend the interactions of consumers within technologies that influence their engagement.

Marketing has long explored various relationships between customers and new technological objects. Various empirical studies have adopted a sensory understanding or an affective relational attitude. Since customers' primary reactions to items are often driven by the sensory application of the items, sensory marketing explored how customers observe items through the various inputs of human senses [3, 36]. In addition, researchers investigated the emotional relationships between customers and objects. These affective relationships occur mainly because of the fact that they are related to owned products, where there exist a (relatively strong) relationship to the product and also relationship for material possessions which influence consumer behavior [19, 22].

To facilitate interactions between customers and AI, marketing managers often select personal service devices to enlarge the understanding of customers' attendance [30]. To illustrate it on the example, in the case of robots, they have a human form, exhibit human characteristics, or imitate human behavior [2]. In a virtual context, mimicking human behavior by chatbots can frequently influence customers that they have interacted with other people [48]. Novak and Hoffman [31] notice an increased agreement in marketing and applied psychology. They state that anthropomorphism is important for understanding how customers apprehend inert objects [24, 47]. Anthropomorphism in this paper refers to the extent to which customers perceive inanimate objects as human, rather than the extent to which organizations create new technologies to appear human. As to Epley et al. stated [7], this recognition stems from "attributing human characteristics to nonhuman technologies."

The noun "anthropomorphism" is also frequently used in brand building marketing literature and primary research on the various marketing communication tools on how marketers use it [12, 39]. Chandler and Schwarz [13] confirmed in their research that anthropomorphism performed a vital role in product decision making and product brand switching. Similarly, several marketing studies have confirmed that anthropomorphic features can lead to a change in attitude toward a specific product brand and its purchase intention [35]. Although anthropomorphism has received considerable attention in the marketing literature, its principal utility is based on the research on technological innovation. In the field of technology, anthropomorphism has received a lot of observation, especially from the actual point of view – robotics.

Some of the notable research that links robotic functions to anthropomorphism is artificiality [2], consciousness and emotions [47], and humanity [9]. Złotowski et al. [50] defined two dimensions of anthropomorphism: uniqueness and human nature. They represented uniquely human qualities with cognitive abilities and human nature with an emotional being that is represented toward an anthropomorphic object, although anthropomorphism is mainly attributed to robotic qualities. Moussawi et al. [27] added that anthropomorphism will be a crucial construct to explore the meaning of personal and digital assistants that can directly influence the customer experience. The successful implementation of anthropomorphism lies in the perception of users and consumers. Moussawi and Koufaris [28] explain perceived anthropomorphism as the degree to which users perceive in-store technology as human. Brahnam [4] emphasizes that, applying anthropomorphic characteristics in chatbot conversations, they gained social acceptance from them. Digital assistants use voice and chat devices with human signals to improve the likability of anthropomorphic characteristics. However, minimal research has explored digital assistant technology and anthropomorphic aspects, particularly from a commercial perspective. This research adds perceived anthropomorphism as an important construct to the study model.

Seeber et al. [42] stated that in addition to functional application as a tool, digital assistants have also become collaborators in the consumer decision-making process. Dellermann et al. [12] supported this phenomenon by pointing out the phenomenon of how these tools came to play an important role in human life. Yang et al. [49] state that smart solutions could improve retail operations and provide personalized customer service. Various literatures support that digital assistants, chatbots, and smart lists are becoming more mature and their interactive system also facilitates human questions.

2 Methodology

The purpose of this paper is to explore customer satisfaction preferences in the context of the household products. On the basis of primary research, we map customer attitudes toward the purchase of products made in Slovakia. We are interested in whether it is important for customers to buy Slovak products and also whether they prefer them when buying food, textiles, and shoes. The research also provides an answer to the question of whether they purposefully choose merchants who offer local and regional products and whether they feel that more foreign products are available in shops than Slovak ones. The aim of the contribution is to make recommendations on how to improve the customer experience by using smart technologies, focusing on the area of interest.

To achieve this goal, representative quantitative research was carried out in the form of a survey among Slovak consumers, while the contribution presented is a partial output of the research. The research was carried out using an online questionnaire in April 2022 and 1000 Slovak consumers aged 18 and over participated in it. The questions concerned the attitude of consumers toward the purchase of domestic products and their attitudes toward the perception of themselves as informed consumers. As part of the contribution, we focused on the evaluation of the following selected statements:

- It is important to me to buy Slovak products.
- I purposely choose merchants that offer regional and local products.
- There are more foreign products available in stores than Slovak products.
- When buying food, I prefer products made in Slovakia.
- When buying shoes and textiles, I prefer foreign products.

The questions in the questionnaire were mandatory, while respondents expressed their level of agreement with the given statements by marking one of the answers on a 5-point Likert scale from strongly agreeing to completely disagree. The research results were evaluated using descriptive statistics. The questionnaire also included classification questions that characterize the respondents from the point of view of sociodemographic characteristics. The sample of respondents fulfills representativeness with respect to gender, age, education, region, and size of residence. A detailed description of the sample based on sociodemographic parameters can be found in the Appendix of the paper.

3 Discussion

Based on the subsequent graphically processed data, it is possible to identify what are the main customer experience preferences in retail stores. We also try to predict through which smart solutions it is possible to achieve improvement in this area and thus offer customers a better shopping experience or directly influence their preferences with a focus on a specific product.

In the research, we mainly focused on customer preferences between traditional Slovak products or competing (supplementary) products. At first, we were interested in whether it is important to customers from which country the given product comes from or if they focus primarily on domestic brands of products that they intend to buy (Fig. 5.1).

Customer affinity and their preferences in this area directly evoke the impression that most customers fall into the category of typical purchasing behavior, where they focus primarily on domestic products. For most customers, domestic products act as a certain standard of product quality and processing. However, 19.70% of respondents show low customer preferences that focus on the domestic origin of the product. For these customers, a foreign brand of products automatically invokes the quality of the product.

Furthermore, we identified whether customers specifically focus on retailers that offer domestic or regional products in their range (Fig. 5.2).

In this direction, the prevalence of a significant part of customer preferences prevails with a focus on merchants offering regional and local products. Nearly

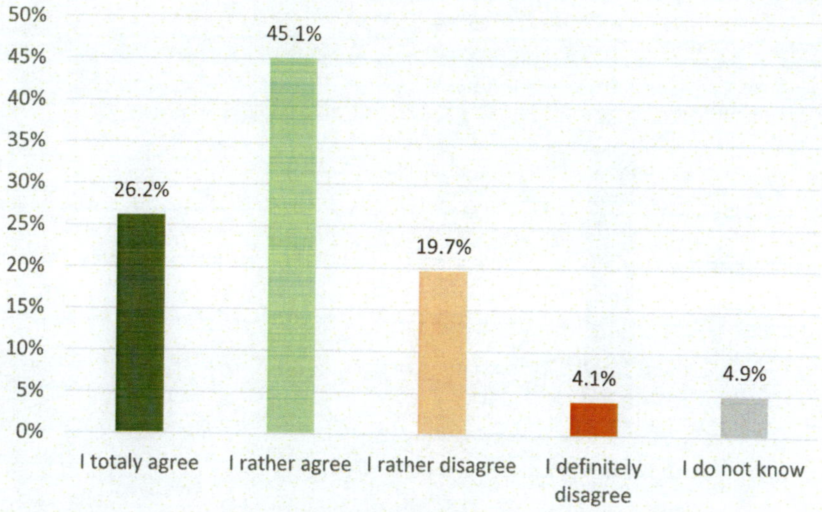

Fig. 5.1 It is important to me to buy Slovak products

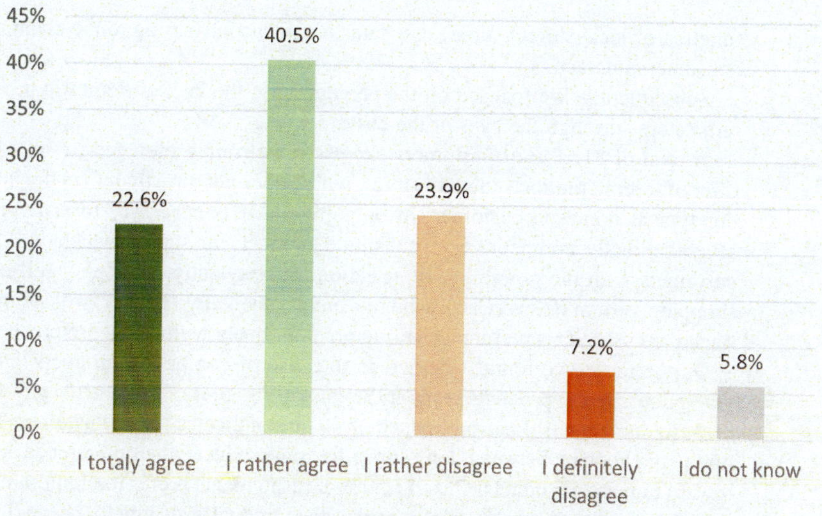

Fig. 5.2 I purposely choose merchants that offer regional and local products

63.10% of respondents prefer this kind of assortment. However, a smaller increase (7.30%) also occurred among customers with medium to low preferences focusing on regional or local products. In this area, it is possible to improve the customer experience through a voice assistant that could help in choosing a store that offers regional products (or indirectly influence the customer with the automatic alerting

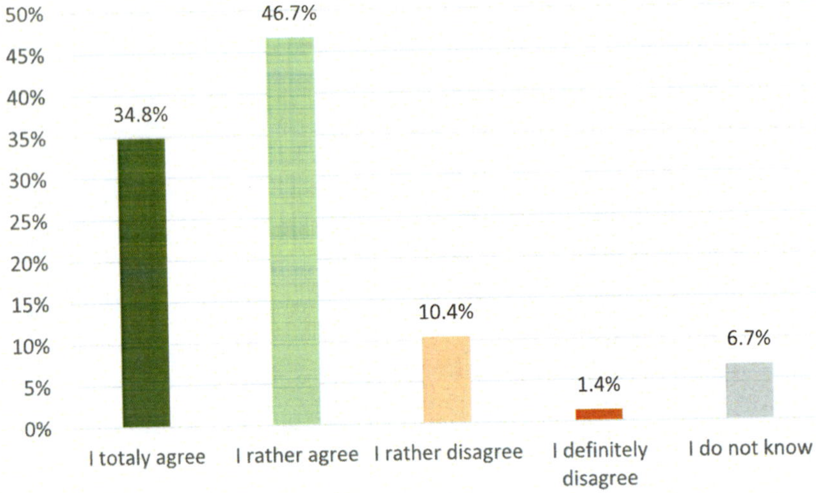

Fig. 5.3 There are more foreign products available in stores than Slovak products

function of local stores). Voice assistants are also contributing to the changing way consumers interact.

Subsequently, we focused on the perception of the lack of domestic products in retail stores through the eyes of the customer (Fig. 5.3).

A total of 81.50% of customers expressed a strong preference for an excessive offer of foreign products in retail stores, which does not directly reflect the interest of our typical domestic customer. In our opinion, it is necessary to start providing several directly competitive or substitute products in stores, which will provide customers with the possibility of selection. Subsequently, through a self-checking shopping system (CleverCart), which is environmentally friendly, we can indirectly track the customer's preference and remove products with a low preference.

Furthermore, we found whether, in the case of competing products, customers give preference to products made in Slovakia (Fig. 5.4).

A significant customer preference of 76.50% indicates that if customers have the option of choosing between domestic or foreign products, their preference is directed toward domestic products. Only 17.50% of respondents prefer foreign products, and 6.00% of respondents do not consider the product's origin to be an important selection factor when purchasing. In this area, we recommend a smart list as a means of improving the customer experience, which can provide customers with information about the origin of a given product in advance and thus simplify the selection process or indirectly influence their choice.

At the end of the research, we were interested in the customer's preference in the case of purchasing textiles and footwear, with a focus on foreign brands (Fig. 5.5).

A total of 12.60 % of customers expressed strong preference and orientation toward foreign brands in the field of footwear and textile products. Furthermore,

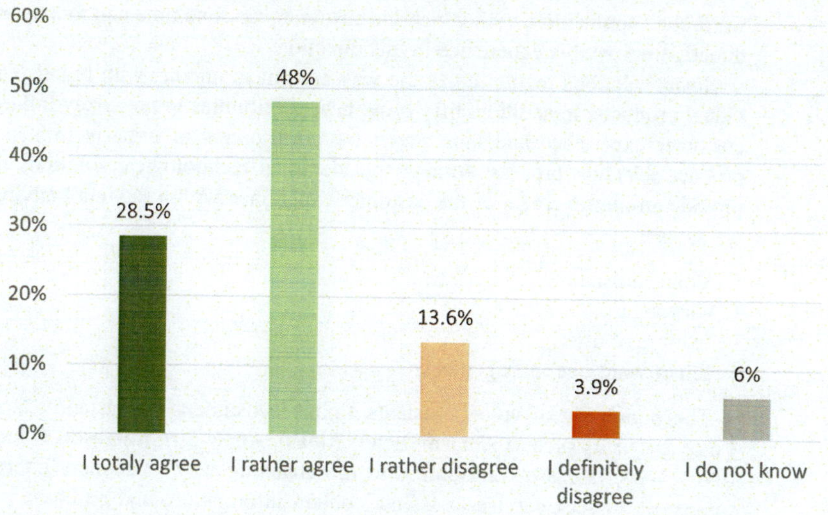

Fig. 5.4 When buying food, I prefer products made in Slovakia

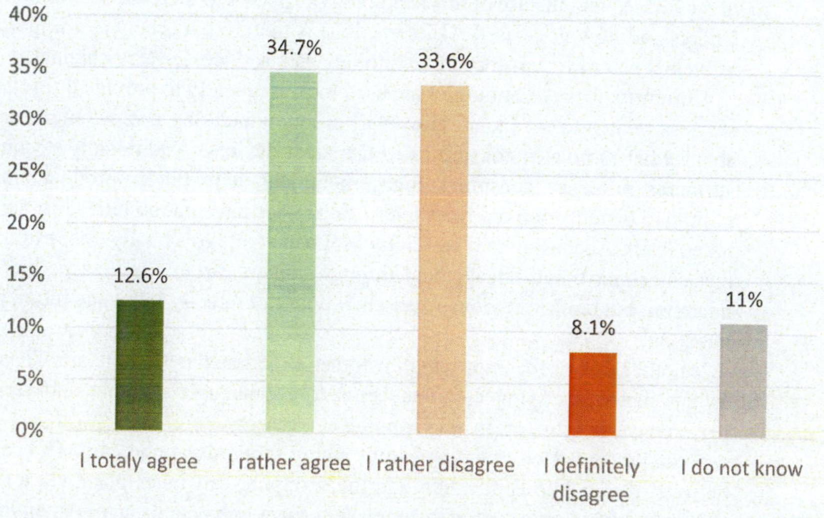

Fig. 5.5 When buying shoes and textiles, I prefer foreign products

34.70 % marked their preference for foreign products as medium and 33.60 % rather preferred Slovak products. Despite the lower representation of Slovak products in this category on the market and their consequently higher price level, we can observe that customers are interested in such products. In this area, we offer as a smart solution that has the potential to improve the customer experience the "smart wardrobe" application, with which the customer can save time and money and thus directly improve his experience in this direction.

Smart solutions revolutionize the way customers interact with brands and individual products. In addition, they bring new possibilities to positively influence the customer experience and thus attract more customers or unfairly influence their product selection process. Among the identified technological solutions that can provide an improvement of the customer experience within the retail environment, we advise:

- Voice assistant
- CleverCart
- Intelligent lists
- Smart wardrobe management

The popularity of voice assistants among consumers is constantly increasing. Voice assistants such as Amazon Echo (Alexa), Apple Siri, Microsoft Cortana, and Google Assistant are contributing to the changing way consumers interact with companies to seek service assistance, obtain information, and purchase products. Accenture [1] reports that 3 out of 10 consumers talk to their assistant more than their family. From this point of view, they can act as a critical driver of service innovation in the future, with the use of this technology expected to increase by 1000% by 2023. Understanding voice-based AI interactions is therefore a timely and important area as brands enter this territory of technology as a service delivery channel.

Currently, these technologies are used most frequently to provide information and services related to the brand. However, despite increasing use, we understand little about what motivates consumers to use such devices. Additionally, technological attributes influence consumer engagement with a particular product along with utilitarian benefits derived from interactions with information related to the product brand itself. According to research by McLean et al. [26], finally, the results suggest that consumer brand engagement through voice assistants influences brand usage intention, but unlike previous research, it does not directly influence future purchase intention.

Another option for improving customer experience is CleverCart. CleverCart is an automated shopping cart that connects to the user's phone and serves as a personal cashier for customers shopping at a grocery store. The CleverCart shopping cart can be connected to the phone through a shopping application. The connection between the mobile application and the shopping cart takes place via a QR code. Subsequently, users can scan the items they want to buy and can put them in the cart; the weighing sensors installed in the cart will check whether the weight of the scanned product and the weight of the product placed in the cart are identical. The app also has features such as a "shopping list" that reminds users of what items need to be purchased, and when customers are done, they can easily choose an electronic payment method to pay. The purpose of CleverCart is to minimize human interaction in grocery stores and speed up the overall payment process.

Another smart solution that can positively influence customer experience is the smart list. Currently, many customers lack the necessary financial knowledge and experience. Through this platform, they can manage their shopping lists, save more on their purchases, and provide them with a seamless shopping experience in physical stores as well as detailed product information. A smart shopping list app, it also aims to encourage customers to build a healthier financial life. Through a user-centered design approach and features such as smart shopping suggestions, budgeting options, the ability to connect with local stores, and a platform for sharing lists, this kind of smart solution has great potential to directly impact the customer experience in the future.

When purchasing shoes and textiles, there are also several alternatives to focus on improving the customer experience. One of them is through the "smart wardrobe" application. Although clothing does not directly define a personality, it certainly helps build character. Therefore, it is quite important for every person to have a wardrobe that they can rely on. The app aims to design a fast-reactive smart wardrobe that can be designed to save money and time while shopping. A smart wardrobe can, for example, notify the user of what clothes or objects are in the wardrobe. This scanning application captures an image of clothes or accessories present in the wardrobe and recognizes the list of items. For individual products, it is also possible to record information about their origin and materials. The captured image with a list of items in the wardrobe is then sent to the application. This helps users avoid buying repeat items and splurging on unnecessary items, while allowing the user to purchase any matching material or accessory within the range. This application is accessible to users anytime, anywhere.

4 Conclusions

Based on the quantitative data, it is possible to find out what the customer experience preferences are in retail stores. We predicted through which smart solutions it is possible to achieve improvement in this area and thus offer customers a better shopping experience or directly influence their preferences with a focus on a specific product. Customer affinity and their preferences in this area directly evoke the impression that most customers fall into the category of typical purchasing behavior, where they focus primarily on domestic products. For most customers, domestic products act as a certain standard of product quality and processing. It is possible to improve the customer experience through a voice assistant that could help in choosing a store that offers regional products (or indirectly influence the customer with the automatic alerting function of local stores). Voice assistants are also contributing to the changing way consumers interact. There is express a strong preference for an excessive offer of foreign products in retail stores, which does not directly reflect the interest of our typical domestic customer. In our opinion, it is necessary to start providing several directly competitive or substitute products in stores, which will provide customers with the possibility of selection. Subsequently, through a self-checking shopping system (CleverCart), which is environmentally friendly, we can indirectly track the customer's preference and remove products with a low preference. Smart solutions revolutionize the way customers interact with brands and individual products. In addition, they bring new possibilities to positively influence the customer experience and thus attract more customers or unfairly influence their product selection process.

The future direction of the topic is to further analyze the smart solutions in the marketing activities, especially those, which are connected to the consumer experience, e.g., consumer care system, creating customer experience. Another direction is to not only include selected product categories but also include the whole consumer basket and the role of smart solutions in the customer buying behavior.

Acknowledgments/Funding This research was funded by the Scientific Grant Agency of the Ministry of Education, Science, Research, and Sport of the Slovak Republic, grant registration number 1/0737/20 – Consumer Literacy and Intergenerational Changes in Consumer Preferences when Purchasing Slovak Products.

Appendix

		Absolute frequency	Relative frequency (%)
Sex	Male	494	49.40
	Female	506	50.60
Age	18–29 years	181	18.10
	30–39 years	198	19.80
	40–49 years	196	19.60
	50–59 years	158	15.80
	More than 60 years	267	26.70
Education	Basic degree	56	5.60
	Secondary vocational education without high school diploma	407	40.70
	Complete secondary education with high school diploma	367	36.70
	University degree	170	17.00
Region	Bratislava region	115	11.50
	Trnava region	104	10.40
	Nitra region	113	11.30
	Trenčín region	129	12.90
	Banská Bystrica region	122	12.20
	Žilina region	129	12.90
	Prešov region	145	14.50
	Košice region	143	14.30
Number of inhabitants	Less than 4999 inhabitants	453	45.30
	5000–19,999 inhabitants	178	17.80
	20,000–99,999 inhabitants	249	24.90
	More than 100,000 inhabitants	120	12.00

(continued)

		Absolute frequency	Relative frequency (%)
Employment	Full time	427	42.70
	Part time	46	4.60
	Maternity/parental leave	38	3.80
	In household	21	2.10
	Student and employee	46	4.60
	Student and no employment	25	2.50
	Retired person	266	26.60
	Entrepreneur	55	5.50
	Unemployed	76	7.60

References

1. Accenture. (2018). *Many voice assistant users have 'trust issues' with their device*. Retrieved from: https://www.independent.co.uk/news/uk/home-news/voice-assistant-amazon-echo-google-home-trustartificial-intelligence-a8534556.html
2. Bartneck, C., Kulić, D., Croft, E., & Zoghbi, S. (2009). Measurement instruments for the anthropomorphism, animacy, likeability, perceived intelligence, and perceived safety of robots. *International Journal of Social Robotics, 1*(1), 71–81.
3. Bosmans, A. (2006). Scents and sensibility: When do (in) congruent ambient scents influence product evaluations? *Journal of Marketing, 70*(3), 32–43.
4. Brahnam, S. (2009). Building character for artificial conversational agents: Ethos, ethics, believability, and credibility. *PsychNology Journal, 7*(1), 9–47.
5. Collier, J. E., & Bienstock, C. C. (2006). Measuring service quality in E-retailing. *Journal of Service Research, 8*(3), 260–275.
6. Eisingerich, A. B., & Bell, S. J. (2008). Perceived service quality and customer trust: Does enhancing customers' service knowledge matter? *Journal of Service Research, 10*(3), 256–268.
7. Epley, N., Waytz, A., & Cacioppo, J. T. (2007). On seeing human: A three-factor theory of anthropomorphism. *Psychological Review, 114*(4), 864–886.
8. Gartner. (2020). *Drive growth in times of disruption*. https://www.gartner.com/en
9. Haslam, N., Loughnan, S., Kashima, Y., & Bain, P. (2008). Attributing and denying humanness to others. *European Review of Social Psychology, 19*(1), 55–85.
10. Hollebeek, L. D., Glynn, M. S., & Brodie, R. J. (2014). Consumer brand engagement in social media: Conceptualization, scale development and validation. *Journal of Interactive Marketing, 28*(2), 149–165.
11. Homburg, C., Jozié, D., & Kuehnl, C. (2015). Customer experience management: Toward implementing an evolving marketing concept. *Journal of the Academy of Marketing Science, 45*(3), 377–401.
12. Hudson, S., Huang, L., Roth, M. S., & Madden, T. J. (2016). The influence of social media interactions on consumer–brand relationships: A three-country study of brand perceptions and marketing behaviors. *International Journal of Research in Marketing, 33*(1), 27–41.
13. Chandler, J., & Schwarz, N. (2010). Use does not wear ragged the fabric of friendship: Thinking of objects as alive makes people less willing to replace them. *Journal of Consumer Psychology, 20*(2), 138–145.

14. Chattaraman, V., Kwon, W. S., Gilbert, J. E., & Ross, K. (2019). Should AI-based, conversational digital assistants employ social-or task-oriented interaction style? A task-competency and reciprocity perspective for older adults. *Computers in Human Behavior, 90*, 315–330.
15. Cheatham, B., Javanmardian, K., & Samandari, H. (2019). *Confronting the risks of artificial intelligence*. https://www.mckinsey.com/business-functions/mckinsey-analytics/our- insights/confronting-the-risks-of-artificial-intelligence.
16. Choi, B., Lee, C., Lee, H., & Subramani, M. (2004). Effects of web retail service quality and product categories on consumer behavior: A research model and empirical exploration 37th annual Hawaii international conference on system sciences, 2004. *Proceedings of the, IEEE*, p. 10.
17. Irfan, M. (2020). *Artificial intelligence and the future of web design*. https://usabilitygeek.com/artificial-intelligence-and-the-future-of-web-design/
18. Keiningham, T., Ball, J., Benoit, S., Bruce, H. L., Buoye, A., Dzenkovska, J., et al. (2017). The interplay of customer experience and commitment. *Journal of Services Marketing, 31*(2), 148–160.
19. Kleine, S. S., & Baker, S. M. (2004). An integrative review of material possession attachment. *Academy of Marketing Science Review, 1*(1), 1–39.
20. Ladhari, R., Souiden, N., & Dufour, B. (2017). The role of emotions in utilitarian service settings: The effects of emotional satisfaction on product perception and behavioral intentions. *Journal of Retailing and Consumer Services, 34*(2017), 10–18.
21. Lam, S. Y. (2001). The effects of store environment on shopping behaviors: A critical review. In M. C. Gilly & J. Meyers-Levy (Eds.), *Advances in consumer research* (Vol. 28, pp. 190–197). Association for Consumer Research.
22. Lastovicka, J. L., & Sirianni, N. J. (2011). Truly, madly, deeply: Consumers in the throes of material possession love. *Journal of Consumer Research, 38*(2), 323–342.
23. Lemon, K. N., & Verhoef, P. C. (2016). Understanding customer experience throughout the customer journey. *Journal of Marketing, 80*, 69–96.
24. MacInnis, D. J., & Folkes, V. S. (2017). Humanizing brands: When brands seem to be like me, part of me, and in a relationship with me. *Journal of Consumer Psychology, 27*(3), 355–374.
25. McColl-Kennedy, J. R., Gustafsson, A., Jaakkola, E., Klaus, P., Radnor, Z. J., Perks, H., & Friman, M. (2015). Fresh perspectives on customer experience. *Journal of Services Marketing, 29*(6–7), 430–435.
26. McLean, G., Osei-Frimpong, K., & Barhorst, J. (2021). Alexa, do voice assistants influence consumer brand engagement? – Examining the role of AI powered voice assistants in influencing consumer brand engagement. *Journal of Business Research, 124*, 312–328. [124]. https://doi.org/10.1016/j.jbusres.2020.11.045
27. Moussawi, S., Koufaris, M., & Benbunan-Fich, R. (2020). How perceptions of intelligence and anthropomorphism affect adoption of personal intelligent agents. *Electron Markets, 31*, 343. https://doi.org/10.1007/s12525-020-00411-w
28. Moussawi, S., & Koufaris, M. (2019). Perceived intelligence and perceived anthropomorphism of personal intelligent agents: Scale development and validation. In *Paper presented at the Hawaii international conference on system sciences (HICSS)*. HI. http://hdl.handle.net/10125/59452
29. Newman, D. (2019). *5 ways AI is transforming the customer experience*. https://www.forbes.com/sites/danielnewman/2019/04/16/5-ways-ai-is-transforming-the-customer-experience/#49e17a31465a
30. Niemelä, M., Heikkilä, P., & Lammi, H. (2017). A social service robot in a shopping mall: Expectations of the management, retailers and consumers. In *HRI '17: ACM/IEEE international conference on human-robot interaction*. Association for Computing Machinery.
31. Novak, T. P., & Hoffman, D. L. (2019). Relationship journeys in the internet of things: A new framework for understanding interactions between consumers and smart objects. *Journal of the Academy of Marketing Science, 47*(2), 216–237.

32. Oh, L. B., Teo, H. H., & Sambamurthy, V. (2012). The effects of retail channel integration through the use of information technologies on firm performance. *Journal of Operations Management, 30*(5), 368–381.
33. Omale, G. (2019). *Improve customer experience with artificial intelligence*. https://www.gartner.com/smarterwithgartner/improve-customer-experience-with-artificial-intelligence/
34. Parasuraman, A., Zeithaml, V. A., & Berry, L. L. (1994). Reassessment of expectations as a comparison standard in measuring service quality: Implications for further research. *Journal of Marketing, 58*(1), 111–124.
35. Payne, C. R., Hyman, M. R., Niculescu, M., & Huhmann, B. A. (2013). Anthropomorphic responses to new-to-market logos. *Journal of Marketing Management, 29*(1–2), 122–140.
36. Peck, J., & Childers, T. L. (2006). If I touch it I have to have it: Individual and environmental influences on impulse purchasing. *Journal of Business Research, 59*(6), 765–769.
37. Prentice, C., Dominique Lopes, S., & Wang, X. (2020). The impact of artificial intelligence and employee service quality on customer satisfaction and loyalty. *Journal of Hospitality Marketing & Management, 29*, 1–18. https://doi.org/10.1080/19368623.2020.1722304
38. Prentice, C., & Kadan, M. (2019). The role of airport service quality in airport and destination choice. *Journal of Retailing and Consumer Services, 47*(2019), 40–48.
39. Rauschnabel, P. A., & Ahuvia, A. C. (2014). You're so lovable: Anthropomorphism and brand love. *Journal of Brand Management, 21*(5), 372–395.
40. Saponaro, M., Le Gal, D., Gao, M., Guisiano, M., & Maniere, I. C. (2018). Challenges and opportunities of artificial intelligence in the fashion world 2018 international conference on intelligent and innovative computing applications (ICONIC). *IEEE, 2018*, 1–5.
41. Saratchandran, V. (2019). *Artificial intelligence (AI): Ways AI is redefining the future of customer service*. https://becominghuman.ai/artificial-intelligence-ai-ways-ai-is-redefining-the-future-of-customer-service-4dc667bfa59
42. Seeber, I., Bittner, E., Briggs, R. O., De Vreede, T., De Vreede, G. J., Elkins, A., Maier, R., Merz, A. B., Oeste-Reiß, S., Randrup, N., & Schwabe, G. (2020). Machines as teammates: A research agenda on AI in team collaboration. *Information & Management, 57*(2), 103–174.
43. Scheidt, S., & Chung, Q. B. (2019). Making a case for speech analytics to improve customer service quality: Vision, implementation, and evaluation. *International Journal of Information Management, 45*(2019), 223–232.
44. *Treasure Data AI vs. Human customer service: Survey data shows when consumers prefer a bot the current state of electronic systems & human interactions* (2019). https://app.hushly.com/runtime/content/hweB8PE1UYgB6oQR
45. van Doorn, J., Mende, M., Noble, S. M., Hulland, J., Ostrom, A. L., Grewal, D., & Petersen, J. A. (2017). Domo arigato Mr. Roboto: Emergence of automated social presence in organizational frontlines and customers' service experiences. *Journal of Service Research, 20*(1), 43–58.
46. Verhoef, P. C., Lemon, K. N., Parasuraman, A., Roggeveen, A., Tsiros, M., & Schlesinger, L. A. (2009). Customer experience creation: Determinants, dynamics, and management strategies. *Journal of Retailing, 85*(1), 31–41.
47. Waytz, A., Heafner, J., & Epley, N. (2014). The mind in the machine: Anthropomorphism increases trust in an autonomous vehicle. *Journal of Experimental Social Psychology, 52*, 113–117.
48. Wünderlich, N. V., & Paluch, S. (2017). A nice and friendly chat with a bot. In *38th international conference on information systems* (pp. 1–11). Association for Information Systems.
49. Yang, G., Ji, G., & Tan, K. H. (2020). Impact of artificial intelligence adoption on online returns policies. *Annals of Operations Research, 308*, 703. https://doi.org/10.1007/s10479-020-03602-y
50. Złotowski, J., Strasser, E., & Bartneck, C. (2014). Dimensions of anthropomorphism: From humanness to humanlikeness. In *2014 9th ACM/IEEE international conference on human–robot interaction (HRI)* (pp. 66–73). IEEE.

Part II
Urban Mobility and IoT

Chapter 6
Construction and Experimental Testing of Innovative Wireless Emission Monitoring System in Small Household of Smart City

Michal Holubčík, Jozef Jandačka, and Miriam Nicolanská

1 Introduction

One of the most critical and current environmental topics of today's era is the reduction of emissions within our environment and on Earth. From the beginning of fuel extraction to the absorption of emissive particles, various technologies were developed in aims of managing the continuously worsening environment health. The most common emissions produced by heating devices via fuel burn are particulate matter (PM), carbon monoxide (CO), and carbon dioxide (CO_2), and substances with significantly larger greenhouse effect are various methane (CH4), nitrous oxides (NO_X), and other sulfuric compounds. Without clear data of how much emissions are being emitted by heat devices and how efficiently fuel is being burned, it is a large understatement to claim that government regulations and mitigations are as effective as possible.

The harmfulness of individual substances to global warming can be described using the coefficient "GWP - Global Warming Potential." The GWP value expresses the equivalent amount of CO_2 released needed to achieve the same warming effect as when 1 kg of a given substance is burned. Emissions are measured relative to a certain concentration of oxygen in the air, thanks to which it is possible to correctly compare the values of emission substances under different conditions (most often to 12% O_2). Table 6.1 summarizes the GWP and health consequences of excessive concentration of substances.

M. Holubčík (✉) · J. Jandačka · M. Nicolanská
University of Žilina, Department of Power Engineering, Žilina, Slovakia
e-mail: michal.holubcik@fstroj.uniza.sk

L. Knapcikova, S. Zohrehvandi (eds.), *9th International Conference on Mobility, IoT and Smart Cities*, EAI/Springer Innovations in Communication and Computing,
https://doi.org/10.1007/978-3-031-28225-6_6

Table 6.1 Overview of emissions their respective GWP and health consequences [1, 2]

Emissive substance	Source of emissions	GWP	Health consequences
PM_{10}	Flow of dust from roads, industrial sources, and fuel burn	–	Easily penetrate the lung tissues, breathing difficulties
$PM_{2,5}$	All types of combustion processes	–	Breathing difficulties
CO_2		1×	Blocking the supply of oxygen to tissues
CH_4		25×	
N_2O	Combustion processes at high temperatures, exhaust gases	273×	Mild to severe inflammation of the bronchi and lungs

The research was initiated with respect to a review of European norms and laws implemented in monitoring and checking boilers. From this research, it was found that a significant number of boilers in Slovakia are burning incorrect fuel and inefficiently and are in average over 14 years old in over 57% population.

Since the goal of the European Union by 2030 to decrease the emissions to 55% lower levels as recorded in the year 1990 and come closer to ideal smart cities, it is essential to build an inexpensive and simple-to-build system to monitor the production of emission in real time.

1.1 Review of Existing Smart Monitoring Systems

In a study originating in India [3], a wireless monitoring network was used to control the temperature and humidity of domestic air conditioning (ventilation, air conditioning, heating). The control system consisted of a microcomputer "Arduino UNO R3" and a sensor "DHT11," which measures the temperature and humidity of the ambient air. The system also consisted of a communication unit "SIM900A GSM Shield," an LCD screen and an adjustable electric fan. The above work suggests that the sensor operates with a sensitivity of ±2 °C to temperature and ±5% RH for humidity, so that the system can measure humidity values between 0 °C and 50 °C and between 20% to 90% RH [4].

According to estimates, more than 52% of the world's population is dependent on indigenous solid fuel heat sources. The World Health Organization (WHO) estimates that 1.5 million premature deaths per year are directly attributable to indoor air pollution from the use of solid fuels in households [5]. The study demonstrated the use of smart sensors such as the MQ-7 (measuring CO) in open-faced solid-fuel stoves.

Another study described the impact of soil pollution caused by various variables that can create a severe risk for the population nowadays [6]. The authors used a method where the air was extracted from soil using a modern multifunctioning device. It can successfully support an easier detection of pollution.

1.2 *Of-the-Counter Emission Sensors and Microcomputers*

From the start of the third decade of the twenty-first century, more and more smart and "of-the-counter" sensors were made available. Such sensors are designed such that the change the resistance of the electrical current within the circuit is related to the concentration of specific emission gases. One important factor is the ability to detect so-called additives: "Additives refer to a group of materials that can alter the ash properties of the burned fuel." [7] Notable mentions are the following:

SDS011: Air Quality Sensor

The SDS011 sensor measures the concentration of particulate matter (PM), specifically PM_{10} and $PM_{2,5}$ (Fig. 6.1). The device utilizes an infrared emitting diode and a photosensitive plate with which the amount of blocked light is measured according to particle size. The dimensions (71 × 70 × 23 mm) of the sensor are small, which allows it to be installed in mobile monitoring systems [8]. Table 6.2 presents the critical technical specifications.

The concentration of PM may be also presented in a volume fraction form within the room. This is because as "the effects of the airflow rate upon the cleanliness of the room are expressed as the percentage of particles present in the reference volumes compared to the number of particles in the total volume" [9].

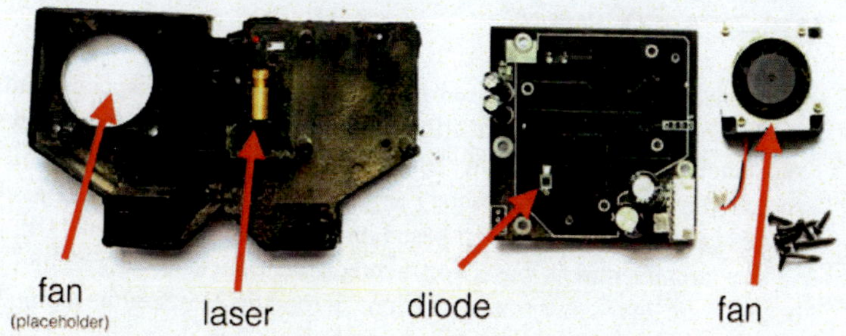

Fig. 6.1 Description of the SDS011 sensor architecture [8]

Table 6.2 Summary of technical information of the SDS011 PM sensor [8]

Technical parameter	Value
Measurement range/speed	0–999 μg/m³/1 s
Maximum/sleep current	100 mA/2 mA
Operating voltage	5 V
Relative error	10%
Temperature range	20–50 °C
Cost in European market	23 €/unit

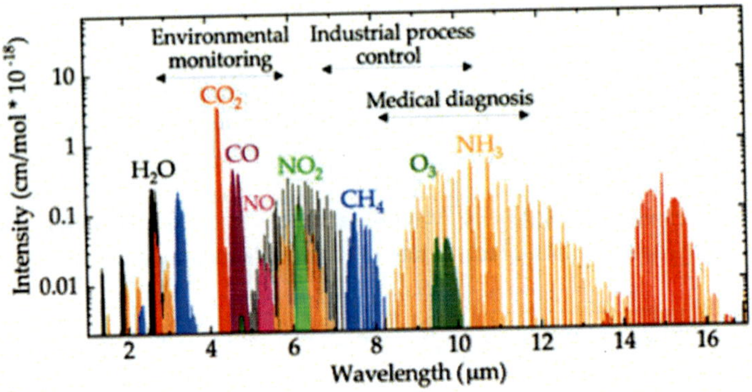

Fig. 6.2 Molecule absorption spectra (H_2O, water; CO_2, carbon dioxide; CO, carbon monoxide; NO, nitric oxide; NO_2, nitrogen dioxide; CH_4, methane; O_3, ozone; NH_3, ammonia)

Table 6.3 Summary of technical parameters of the SEN0332 oxygen sensor [12]

Technical parameter	Value
Measurement range	0–25% Vol O_2
Maximum/sleep current	100 mA/2 mA
Operating volage	3.3–5.5 V (DC)
Resolution	0.15 % Vol
Temperature range	20–50 °C
Cost in European market	46 €/unit

MH-Z19B: CO_2 Sensor

The sensor is an infrared-based gas module using non-dispersive infrared (NDIR) phenomenon in aims of measuring the presence of CO_2 in the air. In short, specific substances absorb certain wavelengths of light, and it's possible to discern based on so-called absorption spectroscopy (Fig. 6.2). The gas in the sample chamber of the device causes light absorption based on the Beer-Lambert Law [10].

The architecture is based on a precision optical circuit design. Some of the benefits are low-power consumption, automatic temperature compensation, anti-water vapor interference, and high resolution. This sensor would allow to monitor the air quality of rooms of buildings or schools where "the operative temperature range was between 21 and 27 °C, while the CO_2 level 700–1300 ppm" [11].

SEN0322: Lambda Sensor O_2

The so-called "Gravity: I2C" oxygen sensor is based on electrochemical phenomenon in which the concentration of oxygen is related to the number of unburnt hydrocarbons after combustion. This enables to evaluate how close to "ideal combustion" the boiler is operating. Summary of technical parameters of the SEN0332 oxygen sensor are shown in Table 6.3.

Nextion: Human Machine Interface (HMI)

The nextion HMI is, simply said, touch screens that offers fully customizable graphical interface of the measured data. Using a designed "Nextion Editor software," it allows a much quicker build of the emission monitoring system to visualize concentrations of gasses (drag-and-drop components such as buttons, sliders, functions, etc.) [13].

1.3 Telecommunication of Wireless Monitoring System

Data transmission functions on the principle of electromagnetic waves that are specifically modulated to hold certain data or information. The role of the telecommunication subsystem is to transform the measured emission concentrations into basic binary components (zeros and ones), modulate it, and send it toward the transmitter. The transmitter then demodulates the incoming data into the emission concentration values. The entire process is simplified using the flow chart of a two-way communication in Fig. 6.3.

In the current market of microcomputers, there are three most common technologies that reliably deliver data without the need of any wire connection:

1. *Wi-Fi modules* – Modules using wireless Local Area Network (WLAN)
2. *Subscriber Identification Module (SIM) technology* – All telephone devices using modern-day SIM cards that carry an identification number, storing data, and information.
3. *LoRa* – The built-in Wi-Fi modules of microcomputers must have a stable connection to Wi-Fi during measurement. LoRa is a long-range low-power radio transmission technique which is commonly used in sending small bits of information over long distances (SIM technology).

All three telecommunication technologies have their benefits strengths and costs with are summarized in Table 6.4. Several technologies are off-the-shelf that provide simple microcomputer integration.

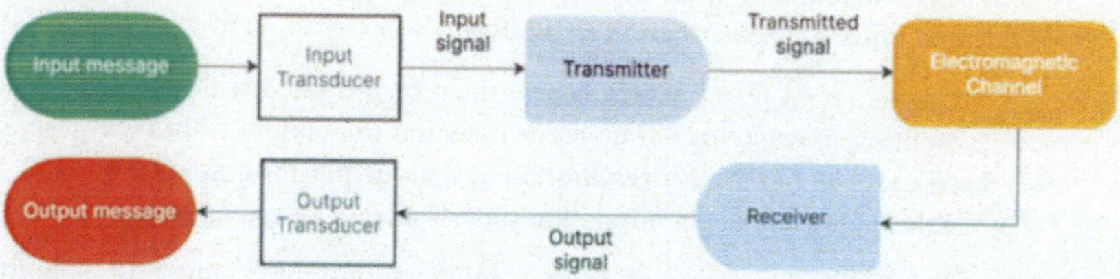

Fig. 6.3 Flowchart diagram of communication network

Table 6.4 Comparison of telecommunication solutions in emission data transfer

Transmission method	Benefits	Costs
Wi-Fi connection	Low-power usage	Necessary proximity to stable Wi-Fi connection
Subscriber Identification Module (SIM)	Unlimited wireless range	Very high-power usage, cost of service
Long-range radio transmission (LoRa)	10–40 km range without Wi-Fi connection	None

Fig. 6.4 Illustration of emission monitoring IoT platform (real-time data)

1.4 *Adafruit IoT Platform and ESP8266 Feather Microcomputer*

The Adafruit IoT platform provides user-friendly data propagation on an online server (Fig. 6.4). The platform provides data feed, visual representation of sensor readings in real time with sharing. The following steps must be followed to connect a "smart emission sensor" to such platform [14]:

1. Create a *data feed* – Every sensor must be provided with one data feed.
2. Define a *specific tag* – The tag is essential to connect within the program.
3. Select *type of data representation* – Various pre-programmed widgets are available such as a gauge, a function plot or a color progress map.

The ESP8266 range of microcomputers provide a low-cost solution of the electronic management system. Such mini controllers range from only 4 € in the

European market with 32 MB drive and 21 pin connections. In specific, the Adafruit Feather ESP8266 allows for simple programming within the Arduino IDE environment along with automatic battery charging via USB connection [15].

1.5 Aim of the Article

The article is a continuation of the design of a wireless emission monitoring system intended for use of nonindustrial heating appliances (household boilers, stoves, etc.).

The key goal of this paper is to provide a description and overall results of the experimental testing of two prototypes of a wireless monitoring device. Both were built to advance air monitoring to a higher level, wirelessly transmitting the data from monitoring into portable smart devices such as computers, tablets, or mobile phones. The novelty of this article is monitoring flue gases produced by heat sources in real-time and wirelessly transmitting data with no presence of an expert.

Additionally, the article aims to present future research and detailed described steps that could help to upgrade the current prototypes and enhance the devices for real-life usage.

2 Construction of Monitoring System Prototypes

During the design phase, two prototypes were built with specific aims. The aim of the first prototype was to demonstrate a "proof of concept" of wirelessly transmitting data from multiple smart sensors simultaneously. The aim of the second prototype was to demonstrate the cooling of the flue gas via the sampling tube, without any damage to the structure and electronics, and to measure emissions produced from a solid-fuel boiler in real time.

2.1 Indoor Prototype

The indoor prototype is aimed to *monitor and measure emissions of indoor rooms, households at room temperatures*. It is shown in Fig. 6.5 that it is divided into two main parts. The first part, or "electronic chamber," contains all electronic devices, i.e., batteries, computer technology (Arduino Uno), and GSM module. In the second, the so-called sensor chamber, emission sensors, and a thermocouple are glued, which are connected by cables to a microcomputer. The fan located on top of the structure creates the necessary air flow through the sensors by sucking in air from the outside and thus distributing it to the sensor chamber, where it passes through each sensor in turn.

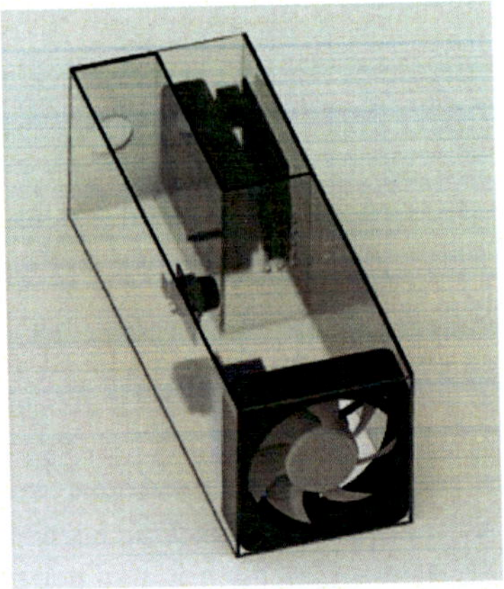

Fig. 6.5 CAD model of the indoor emission monitoring prototype

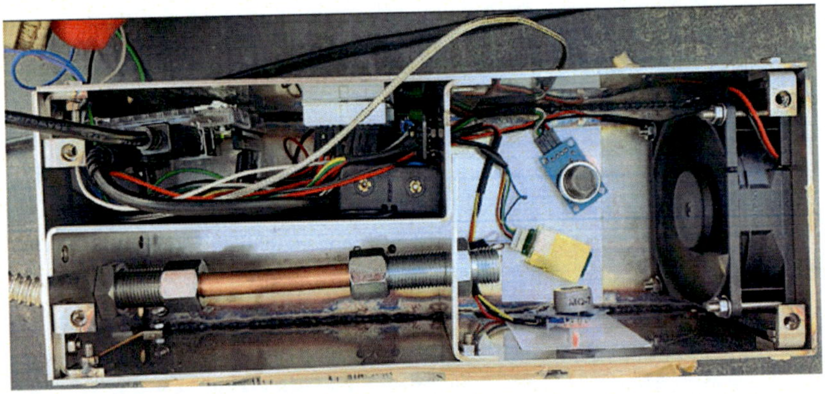

Fig. 6.6 Inside architecture of the monitoring prototype from metal (including battery pack)

To date, all data from similar devices has been shared using Wi-Fi technology, but the signal of which is not guaranteed in the chimneys of various buildings, and the connection would not be stable. Therefore, the second prototype made of steel material (and the following prototypes) will include a GSM module with a working SIM card and connection to an Arduino computer.

2.2 Outdoor Prototype

The second prototype is intended to *analyze flue gasses emitted by boilers during operation*. Figure 6.6 depicts three important segments: the sensor chamber, the cooling chamber, and the electronics chamber. As can be seen also in this figure, the individual components are fixed in position onto the case using either screws, glue, or simple heat-resistant tape.

3 Experimental Testing and Results

Design of monitoring prototypes are required to pass certain regulations, norms, and technical specifications. One of the most important requirements is the fact that the data are gathered accurately enough to reflect the quality of the fuel burn. Another requirement is in relation to data processing; for instance, no data are lost during the wireless transfer (even in worse weather conditions). Lastly, arguably the most critical and top-level requirement is that the device is safe to use in the proximity of the chimney at high flue gas temperatures and harsh environment and in no way shall the monitoring system hinder the chimney of extracting dangerous and contagious smoke out of the system.

The prototypes were tested in various experimental setups. Overall, three experiments were performed. The points summarize the experiment setup:

Experiment n. 1 (*sensor testing*) – Testing and evaluation of the individual electronic sensors in a controlled environment. The aim is to show how the sensors react to an emission source (candle fire, smoke, dusty room).

Experiment n. 2 (*system test*) – System compatibility test within which the device and the sensors are connected via USB. The aim is to show if the sensors are compatible with each other and to show any deviations or interference.

Experiment n. 3 (*Full scale wireless transmission test*) – Testing behavior of the lithium-ion batteries along with the signal strength from the chimney vicinity. The aim of the experiment is to show actual flue gas sampling and cooling.

3.1 Sensor Testing

The test of dust sensors required a space in which forced air pollution by dust was caused. The sensor GP2Y1010AU0F from Sencor evaluates only PM2.5 particles, and during the experiment, the values derived from it were very irregular, therefore untrustworthy in nature. For this reason, the sensor was excluded from the list of potentially usable sensors and was replaced by a second type of sensor (Fig. 6.7).

The carbon monoxide (CO) concentration sensor was exposed to fresh air for 10 seconds, then it was placed under a small, closed container, in which the candle was extinguished (time interval 30 seconds). As can be seen in the curve below, the sensor displayed a value in space from a time of 10–20 seconds from a value of 8 ppm to an increased value of about 14 ppm (a typical concentration of flue gases in a chimney) (Fig. 6.8).

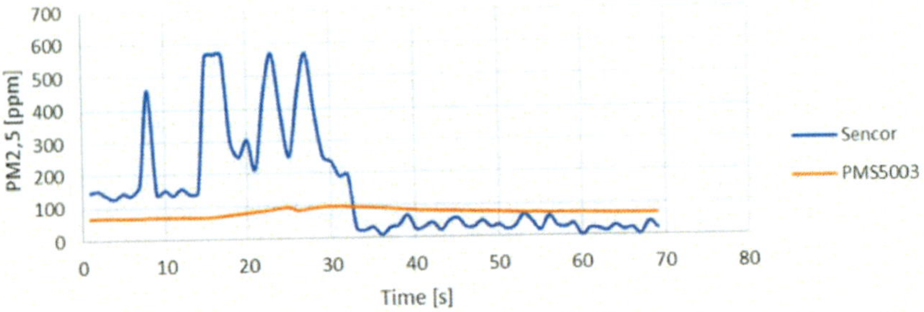

Fig. 6.7 Comparison of two PM sensors: static test

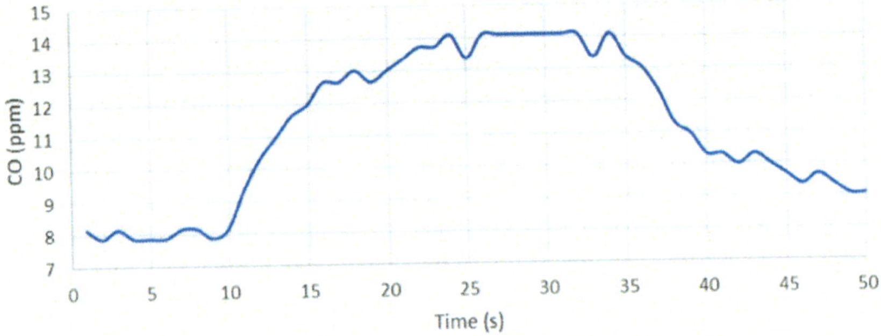

Fig. 6.8 Testing of the MQ-7 sensor of carbon monoxide concentrations

Table 6.5 Average reading of testing environment

Parameter	Average	Standard deviation
Ambient temperature	21.85 °C	0.31 °C
PM10	662.98 ppl	14.04 ppl
PM5	586.50 ppl	11.88 ppl
PM2,5	64.35 ppl	6.93 ppl
CO	1.64 ppm	0.04 ppm
VOC	22.31 ppm	1.4 ppm

3.2 *Experimental Results Within Solid-Fuel Boiler Test*

The second test was a solid-fuel boiler in which three different fuel types were compared based on quality: dry wood, moist paper, and grass/leaves. The goal of the experimental setup was to prove that the sensors will indicate and successfully measure the increase in smoke production as worse combustion processes shall take place. Firstly, it was essential to measure the environmental nominal values of air quality around the test vicinity.

Table 6.5 presents the average value and the standard deviation of a 60 seconds reading of both particulate matter (PM), carbon monoxide (CO), and volatile organic

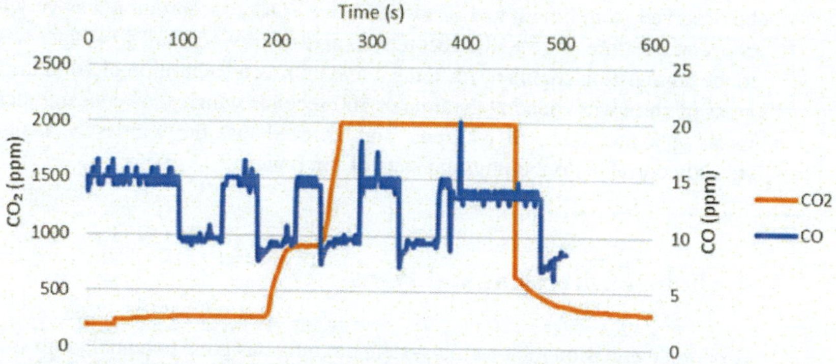

Fig. 6.9 Time progression of the CO and CO2 concentrations during fuel burn

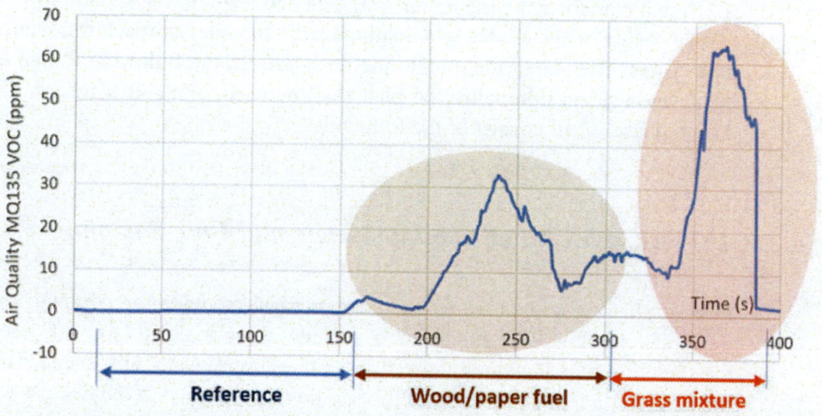

Fig. 6.10 Fuel quality detection with VOC measurement (red = dangerous; brown = mild)

gases (VOC). The results of this experiment show insignificant fluctuation of the research parameters, therefore, demonstrating stability.

The prototype was placed on the side of the chimney from which the sampling tube was inserted in the chimney. The device was turned on, the boiler burning wood at a steady rate, and the sampling probe was inserted in the center of the chimney opening in a depth of approximately 30 cm. Figure 6.9 depicts the time progression of the CO and CO2 concentrations of gas within the air flow. The MQ-7 sensor shows faulty behavior as the value jumps in a step-function-like manner, which is later associated with logistics damage. The other MH-Z19 depicted a clear pattern reaching a maximum reading of 2000 pm, with a climb rate of approximately 60 seconds.

Figure 6.10 depicts the reading of the VOC sensor during burn time. The results showed a clear indication of inappropriate and ineffective fuel burn within the system, completely wirelessly via Wi-Fi into the database "ThingSpeak." As can be observed from time 0 to 150 seconds, the VOC concentration was very low and constant. At time of 150–300 seconds, a mixture of wood and wet paper was added to the combustion chamber, reaching a maximum concentration of 24 ppm. The wet grass mixture was added at the time of 300 seconds which showed a significant gap, reaching a maximum of 64 ppm. This showed that the wirelessly attained data strongly correlate to the physical state of the boiler.

4 Future Directions and Plans

A portable emission monitoring system not only has great potential in measuring the current state but also allows to combine new modular technologies to actively reduce emissions. The design may consider technologies such as electrostatic precipitators (ESP) that could use the input emission data and adapt to the situation.

The design of the device will continue with an agile approach following a rapid prototyping. Regarding the monitoring design itself, the following design development, testing, and optimalization plan shall be used. With such regard, the future design phases shall consist of the following:

4.1 FEM Model of Cooling System and Flue Sampling

First of all, it is imperative to build a numerical model using computational fluid dynamics (CFD). The model will include a normalized size of a chimney, the sampling tube including the ventilator that forces air to flow inside. The aim of the model is to model the temperature and flow distribution of the flue gas within the sampling tube of the monitoring device. Figure 6.11 presents a scheme of the numerical model.

The boundary conditions consist of a constant heat source within the fireplace (bottom of the chimney). The side walls allow heat to escape based on heat conduction of the brick wall (average burn rate of 15–55 kW/m^2) [16]. Both conduction and convection heat transfer shall be modeled from within the sampling tube as outdoor temperature has been showed from experimental data to be a large cooling contributor. A constant pressure source shall be implemented at the end of the monitoring device of a 12 W ventilator (static pressure of approximately 8.64 mm H_2O or 84.75 Pa).

4.2 Experimental Setup with LoRa Communication

The goal of the experiment is to prove the concept and effectiveness of using LoRa communication technology in data transmission. A prototype shall be built around the microcomputer ESP8266 with built-in Wi-Fi module. Multiple sensors shall be

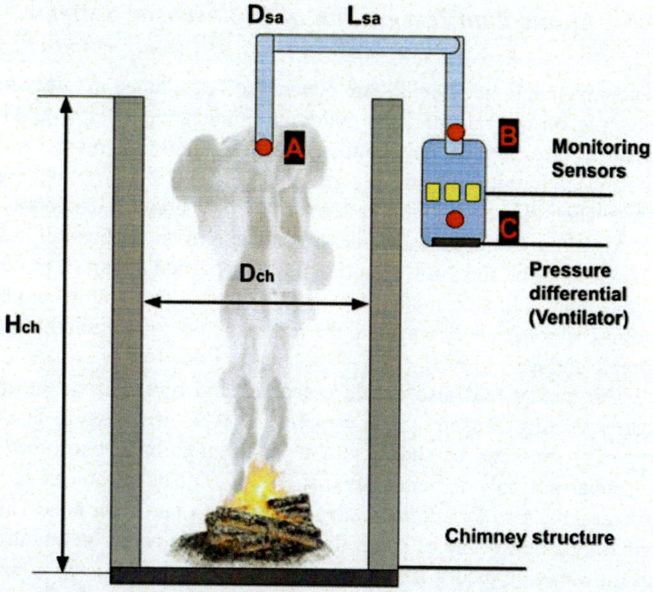

Fig. 6.11 Schematic representation of the numerical heat model

tested individually based on a simple test above a constant emission source (for instance, a candle flame or smoke from cigarette). Additionally, the monitoring system must be evaluated during start-up phase to evaluate the sensor performance.

Intended Electronic Architecture

- ESP8266 – microcomputer
- SX1278 – LoRa receiver/transmitter module
- 3.3 V Ion Lithium battery
- Nextion 2.8′ touch display

Intended Air Quality/Burn Quality Sensors

- *MQ-7* – Sensor measuring concentration of *carbon monoxide* (CO)
- *MH-Z19* – Sensor measuring concentration of *carbon dioxide* (CO_2)
- *DHT-22* – Sensor measuring *air temperature* and *humidity*
- *SEN0322* – Lambda sensor measuring concentration of *oxygen* (O_2)
- *SDS011* – Sensor measuring particulate matter (PM_{10} and $PM_{2,5}$)

Should the sensor show unstable or unreliable readings using 3.3 V connection, the sensor shall be powered using the VIN pin wish 5 V voltage. Additionally, the results shall be compared using the ESP32 microcomputer with similar technical specifications.

4.3 Application Integration of Multisensor System

Since the mentioned devices are planned to be included in larger future testing (for example, on several smart city buildings simultaneously), it would be appropriate to create a corresponding application. This would faithfully display not only the results of devices monitoring but also recommendations and warning signals to the end users when the concentration of emissions rises above a dangerous level.

Also, one of the essential ideas to include in an application is a map where the results could be shown. It would represent a particular part of the smart city and the concentration of emissions produced in the area. This map would be structured in a clear, custom-friendly way, and the representation of measurement results would be simple and easy to read.

The map would also divide outcomes and results from the measurements by corresponding sensors. That way, every type of emission could have its own representation and be shown in a separate graph or leveling meter. This could be essential not only for end users, such as residents of houses or users of heating sources, but also for administrative workers in a particular area. This will offer them not only a better way of controlling a level of air pollution but also a possibility of acknowledging places where the pollution is the greatest (by graphical and numerical representation in a map).

5 Conclusion

The sensor GP2Y1010AU0F developed by Sencor did show accurate readings of particulate matter concentration in a study conducted in South Korea. The study [17] attained a part-to-part repeatability of only 2 μm/m^3. This was achieved using a straightforward short-term noise reduction technique, an auto-fitting calibration for part-to-part repeatability to identify the baseline of the signal that influences the system's performance, as well as a technique for temperature and humidity compensation. This shows that error-reduction techniques must be implemented in future research.

The readings and calibration of the MQ-7 have been conducted in the study [18]. The study implemented best-fit-curve practice during the calibration process. This led to a more accurate model of the relationship between the change in resistance within the sensor and the actual concentration of CO molecules in the air. Therefore, it is highly recommended to manually calibrate the sensor (not relying on automatic calibration) to discern possible outliers and inconsistencies within the setup. Another additional control step would be implemented as: "The measurement is extended by a thermo-gravimetric analysis of samples of the fuels used, supplemented with an analysis of the emitted particles." [19].

Both prototypes successfully measured, monitored, and transferred wireless data onto a IoT platform from which users could continuously monitor the emission production from a remote location. To conclude the overall condition of both experimental devices, it is said that the devices can locate smoke in the flue pipes and can determine the air quality in the vicinity of the given heat sources. None of them were damaged or harmed during the experiments. Hence, they will take part in future research.

The replacement of the existing emission sensors with advanced ones (with a larger monitoring range) is considered and recommended. It is a longer term point of view meant to ensure larger stability, functionality, and reliability. One of the main goals of future research is to acknowledge the device's lifespan, the duration of one charge of the batteries, and the lifespan of each sensor used in the devices.

An ideal direction for the future would be to start the so-called new direction in the acknowledgment of the population to maintain a clean environment and air quality. It could create a new modern way of monitoring the overall health state of the population and an advanced perspective of an intelligent and responsible smart city.

Acknowledgments This publication has been produced with the support of the VEGA 1/0233/19 "Construction modification of the burner for combustion of solid fuels in small heat sources", APVV-21-0452 Impact of using small electrostatic precipitators to reduce particulate pollutants in domestic fuel combustion and KEGA No. 032ŽU-4/2022 Implementation of knowledge about modern ways of reducing environmental burden in the energy use of solid fuels and waste into the pedagogical process.

References

1. EPA. *Understanding global warming potentials*. https://www.epa.gov/ghgemissions/understanding-global-warming-potentials. Accessed 2022/09/10.
2. EPA. *Health and environmental effects of particulate matter PM*. https://www.epa.gov/pm-pollution/health-and-environmental-effects-particulate-matter-pm
3. Shatshikh, Y., Sharapov, A., & Byankin, I. (2017). Analysis of deep heat recovery from flue gases. *Journal of Physics, 891*(5), 012188.
4. Arduino GSM Shield V2. https://www.arduino.cc/en/Main.ArduinoGSMShield. Accessed 2022/05/24.
5. Murulidhar, K. S., Sabari Vignesh, S., Tejaraj, M., & Manikanta, M. R. (2018). Emission control and fuel efficient by using arduino based self-regulating biomass stove. *International Journal of Management Technology And Engineering, 10*(1), 378–391.
6. Skvarekova, E., Tausova, M., Senova, A., Wittenberger, G., & Novakova, J. (2021). Statistical evaluation of quantities measured in the detection of soil air pollution of the environmental burden. *Applied Sciences, 11*(7), 3294.
7. Nosek, R., Werle, S., Borsukiewicz, A., Żelazna, A., & Łagód, G. (2020). Investigation of pellet properties produced from a mix of straw and paper sludge. *Applied Sciences, 10*(16), 5450.
8. AQICN. *The SDS 011 sensor, real time air quality index*. https://aqicn.org/sensor/sds011/. Accessed 09/09/2022.
9. Mičko, P. (2022). Impact of the speed of airflow in a cleanroom on the degree of air pollution. *Applied Sciences, 12*(5).

10. Popa, D., & Udrea, F. (2019). Towards integrated mid-infrared gas sensors. *Sensors, 19*(9).
11. Orman, L. J., Majewski, G., Radek, N., & Pietraszek, J. (2022). Analysis of thermal comfort in intelligent and traditional buildings. *Energies, 15*(18).
12. DRRobot. *SEN0322 oxygen sensor*. https://wiki.dfrobot.com/Gravity_I2C_Oxygen_Sensor_SKU_SEN0322#target_2. Accessed 09/09/2022.
13. Nextion. https://nextion.tech/. Accessed 11/09/2022.
14. Adafruit. *The internet of things for everyone*. https://io.adafruit.com/. Accessed 15/09/2022.
15. Adafruit Feather HUZZAH ESP8266. https://learn.adafruit.com/adafruit-feather-huzzah-esp8266. Accessed 13/10/2022.
16. Tran, W. C., & White, R. H. (1992). Burning rate of solid wood measured in a heat release rate calorimeter. *Fire and Materials, 16*(1), 197–206.
17. Hyuntae, C., & Yunju, B. (2021). Practical particulate matter sensing and accurate calibration system using low-cost commercial sensors. *Sensors, 21*(18).
18. Kobbekaduwa, N., de Mel, W. R., & Oruthota, P. (2021). Calibration and implementation of heat cycle requirement of MQ-7 semiconductor sensor for detection of carbon monoxide concentrations. *Advances in Technology, 1*(2), 377–392.
19. Poláčik, J., Sitek, T., Pospíšil, J., Šnajdárek, L., & Lisý, M. (2019). Emission of fine particles from residential combustion of wood: Comparison of automatic boiler, manual log feed stove and thermo-gravimetric analysis. *Journal of Cleaner Production, 271*(1).

Chapter 7
Optimizing the Resource's Mobility, Transportation, and Networking in the Project Construction Industry, Using a Fractal Method

Esmaeil Cheraghi, Shakib Zohrehvandi, Lucia Knapčíková, Shahram Shadrokh, Roya Soltani, and Mohammad Khalilzadeh

1 Introduction

In case of increasing of projects' number in the company, choosing a completion strategy to supply resources in profit and loss and finishing the work according to the schedule becomes very important. Contracting companies in the country's construction field may have projects in different parts of the country or geographically [6]. When the projects are in different geographical areas, providing the resources

E. Cheraghi · M. Khalilzadeh
Department of Industrial Engineering, Science and Research Branch, Islamic Azad University, Tehran, Iran

S. Zohrehvandi (✉)
Faculty of Manufacturing Technologies with a seat in Prešov, Department of Industrial Engineering and Informatics, The Technical University of Košice, Prešov, Slovak Republic

New Technologies Department, Center for European Studies, Kharazmi University, Tehran, Iran
e-mail: shakib.zohrehvandi@srbiau.ac.com

L. Knapčíková
Faculty of Manufacturing Technologies with a seat in Prešov, Department of Industrial Engineering and Informatics, The Technical University of Košice, Prešov, Slovak Republic

S. Shadrokh
Department of Industrial Engineering, Sharif University of Technology, Tehran, Iran

R. Soltani
New Technologies Department, Center for European Studies, Kharazmi University, Tehran, Iran

Department of Industrial Engineering, Faculty of Engineering, Khatam University, Tehran, Iran

L. Knapcikova, S. Zohrehvandi (eds.), *9th International Conference on Mobility, IoT and Smart Cities*, EAI/Springer Innovations in Communication and Computing,
https://doi.org/10.1007/978-3-031-28225-6_7

needed by the projects will play a key role in the timely completion of the project and the profitability of the project and the company. The available solutions for obtaining these resources or direct purchase from factories by any project or acquisition of mines in the form of purchase or lease and construction of factories and provision of required materials are the company's responsibility. If the volume of work increases the number of projects in the region, it is cost-effective to participate in the preparation and supply of materials. Otherwise, investing in this sector is not profitable.

Liu and Wang [9] presented a flexible model for solving linear scheduling problems involving different objectives and resource allocation tasks. The concept of resource outsourcing is introduced here to improve project performance. According to the research results, considering outsourcing resources not only helps to achieve the diversity of resource supply but also creates a positive impact on the optimal solution by simultaneously reducing time and cost. Cheng et al. [5] developed a Petri net model to transform the information constraints across the information constraint network. A resource management model using an information constraint network is developed to simulate and optimize resource allocation in the design process. The result shows that the proposed approach can achieve the resource management and optimization needed to shorten the development cycle and optimal resource allocation.

In this article, the traditional concept of resource planning is changed. For example, the headquarters manages the resource inventory of the regions, and the regions manage the resource inventory of the projects. Also, the possibility of supplying between group members in the same category is considered to avoid delays and ensure the smooth flow of resources. In general, resources include human resources, materials and machinery, and workshop equipment. This paper aims to propose a fractal resource requirement planning (FRRP) model in project-oriented construction and industrial companies by providing a fractal conceptual framework based on resource and inventory management to minimize resource costs and the smooth flow of materials between members to meet the existing demand. Assuming three levels of headquarters, regions, and projects, each member in the collection is a self-similar structure defined as a fractal.

The innovation of this research is proposing a fractal resource management model in project-oriented construction and industrial companies for the first time with a fractal conceptual framework based on resource and inventory management to minimize resource costs and fluid flow of materials between members to meet existing demand. The research question is: Can the fractal model presented in this research be used and implemented for resource management in project-oriented construction companies?

2 Literature Review

Resource planning is one of the most important competitive and profitable factors in manufacturing and construction industry. To control resources and costs must be used as efficiently as possible. This can be achieved by minimizing the total cost of leased resources subject to the maximum and most efficient use of owned equipment and contracted labor [7]. Rahmanniyay and Yu [11] proposed a new multi-objective mathematical model for developing a project team for multidisciplinary projects under uncertainty. Objective functions optimize cost and merit simultaneously. Based on system dynamics theory, modeling, cost, and environmental benefits of construction and demolition waste resource management under life cycle assessment are proposed by Sun et al. [12]. The results demonstrate that a single policy cannot simultaneously improve environmental benefits, illegal dumping, recycling, and landfill behavior.

Chen et al. [4] presented a scheduling system that can help project managers find near-optimal schedules according to their project objectives and constraints. Zohrehvandi [14] proposed a heuristic algorithm for a power generation project construction to determine the size of project buffer and feeding buffers. This algorithm consists of three phases: (1) determining the duration of activities, (2) demonstrating the size of the project buffer and feeding buffers, and (3) simulating the proposed algorithm. The findings demonstrated that the appropriate number of buffers can be assigned for projects using the proposed algorithm. In another research, Zohrehvandi et al. [15] proposed a project buffer and resource management (PBRM) algorithm to optimize project resources and the size of the project buffer and control of consumption of the project buffer in a renewable project to achieve a more realistic project duration and project completion time. Findings showed that it can be concluded that by using the PBRM algorithm in the construction of this wind turbine project, the duration of the project is 25% less than the planned duration and 29% less than the average completion duration of previous projects.

The idea of fractal companies is an organizational conceptual plan that seeks to achieve a high degree of flexibility to react and quickly adapt to environmental changes by using decentralized and independent organizational units called fractals. Fractal firms are therefore viewed as a temporary set of client-server and board relationships between project managers that interact to diversify product portfolios, obtain economies of scale, and share overhead costs. Canavesio and Martinez [2] proposed a conceptual model for small and medium-sized networks based on the fractal enterprise approach and concepts such as projects, resources, goals, specialized actors, programs, and their relationships. In this work, each fractal management unit is modeled as a project. Peralta and Soltero [10] reviewed publications that develop fractal systems for manufacturing and industries. This review includes contributions published between 1985 (the approximate date of the first works in applied theory in structural engineering) and 2019. The aim was to collect methodologies, strategies, and successful real case studies that can be useful for future studies to apply Industry 4.0 approaches.

Bider et al. [1] proposed a new type of organizational model called the fractal organizational model with methodological support for their design. Karaulova et al. [8] presented a new method for planning and parameter evaluation for the multi-project management case. Using the fractal approach in multi-project management does not require special knowledge of complex system theory. In this research, a fractal is integrated into the production planning framework in this environment. Canavesio et al. [3] analyzed the use of agent-based simulation to illustrate the behaviors that emerge from the interaction of project managers and resource managers when establishing client-server relationships. The results showed that a project-based fractal company model allows virtual and temporary integration between different companies to achieve specific business goals.

According to the literature review that has been done in this research, it was found that little research has been done about the use of fractal methods in the resource management of companies and factories and proposing the FRRP model is for the first time in project-oriented construction and industrial companies. Therefore, this shows the novelty of the proposed FRRP model in this research.

3 Research Methodology

In this research, resources include human resources, materials and machinery, and workshop equipment. This paper aims to propose an FRRP model in project-oriented construction and industrial companies by providing a fractal conceptual framework based on resource and inventory management to minimize resource costs and the smooth flow of materials between members to meet the existing demand. Assuming three levels of headquarters, regions, and projects, each member in the collection is a self-similar structure defined as a fractal. Based on this, the three layers in Figs. 7.1, 7.2, and 7.3 have been created and Fig. 7.4 shows the conceptual model of this research. Figure 7.1 illustrates the first layer (the headquarters) of the proposed FRRP model.

Figure 7.2 shows the second layer (the set of activity areas) of the proposed FRRP model.

Figure 7.3 demonstrates the third layer (the collection of projects of the region) of the proposed FRRP model. Each project can use the facilities and resources available in neighboring projects in the same region. The use of the resources of the projects of a region is to meet the needs again and complete the required resources after the general supply from the regions and the headquarters.

As Fig. 7.3 shows, projects can use the company's headquarters or regions to provide their resources. It has been shown in the studies that the use of the fractal method to optimize the supply chain network increases the speed and reduces the cost of the entire network [13].

Figure 7.4 shows the conceptual model of the proposed research. This conceptual model demonstrates the relations between layers 1, 2, and 3. The central headquarters and the regions are the distributors of resources, and the projects use the resources. The input and output are similar in the three layers above. Organization

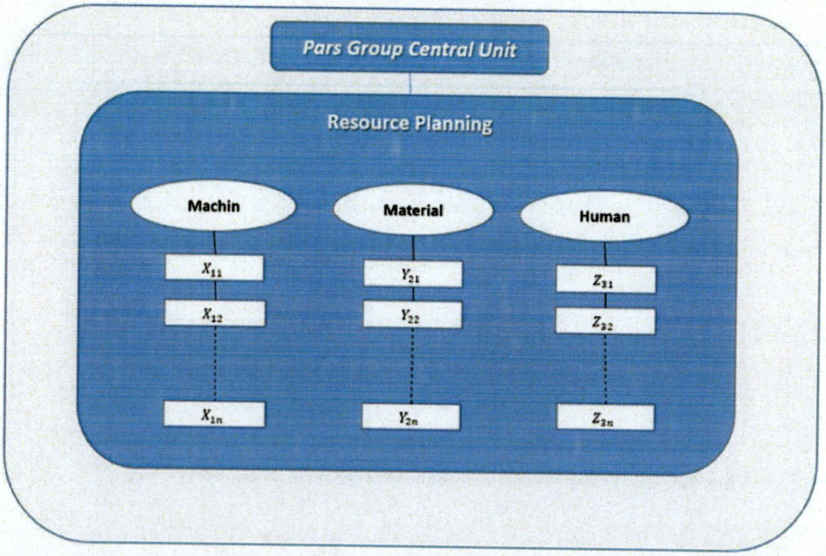

Fig. 7.1 First layer: the headquarters

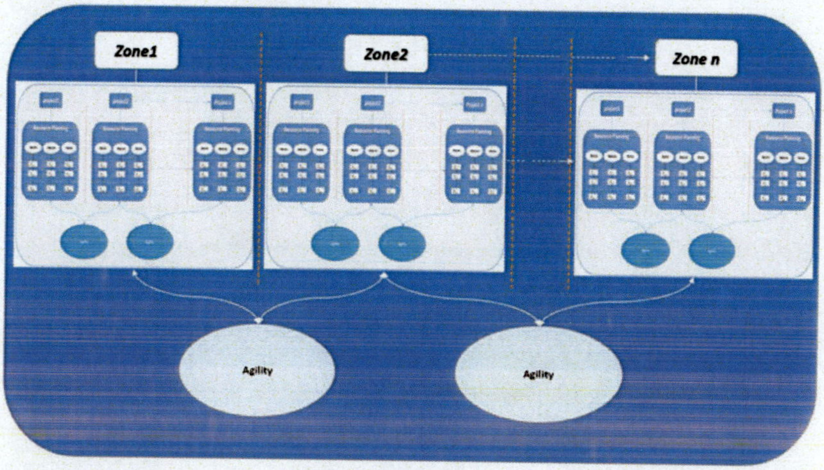

Fig. 7.2 The second layer: the set of activity areas

and optimization can also be done in each layer. Self-organization, self-optimization, goal-oriented, self-similarity, and dynamics are the main characteristics of fractals. If optimization is done in each layer by each member of the network, with the existence of a dedicated network among the members of each layer in the entire company, it is possible to optimize the entire network.

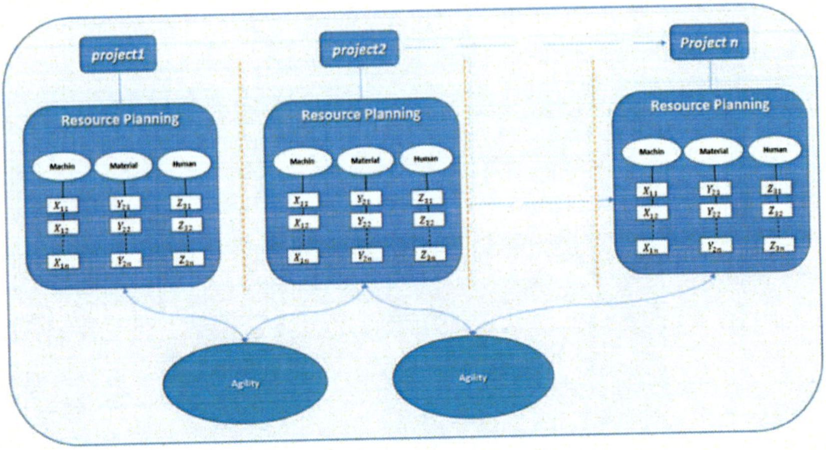

Fig. 7.3 The third layer: the collection of projects of the region

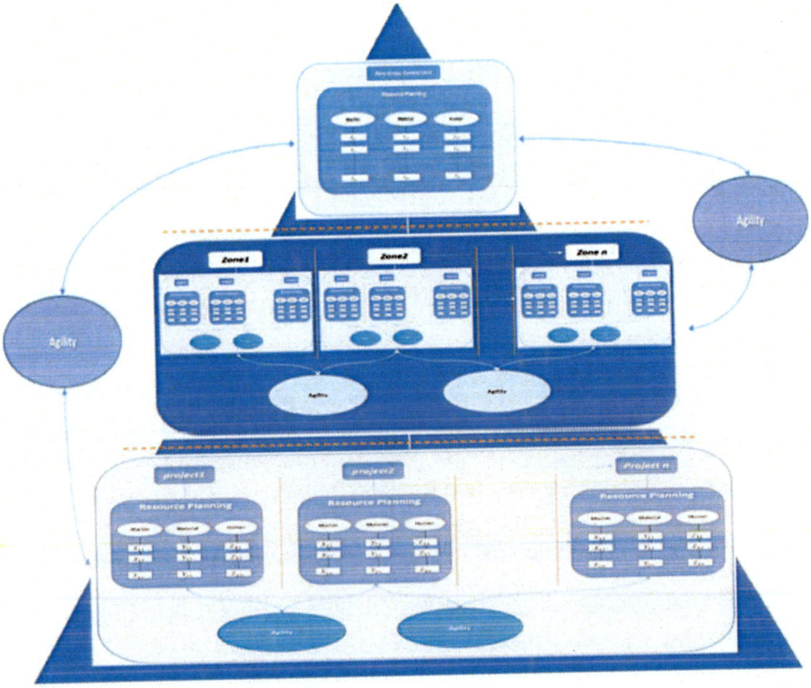

Fig. 7.4 Conceptual model of the research

In this research, to model the research problem based on the conceptual model, the mathematical problem of inventory management has been developed based on the fractal system for purpose of minimizing the costs of transportation and movement of resources.

3.1 Calculation Method Based on the Conceptual Model

If a fractal consists n sub-fractals, the costs of the fractal with an inventory level value q, C_q, include the cost of inventory in the sub-fractal, the cost of the backlog in the sub-fractal, the cost of the shipping, and the cost of the ordering for an inventory level amount of q can be calculated as follows:

$$C_q = \sum_{i=1}^{n} \left\{ E(C_{\text{inv}i}) + E(C_{\text{BC}i}) + C^{\text{f}} + \sum_{i=1}^{n} C_{qi} \quad I\forall i, \right. \tag{7.1}$$

Equation 7.1 shows the fractal cost with replenishment q value. The lower level fractal cost (C_q) can be calculated in a similar way until the completion of the costs of all fractals calculation. Overhead costs are customized depending on the fractal position. The goal of each fractal optimization is concluded by the results of the analysis. The optimization process is first implemented locally in each sub-fractal. After that, the sub-fractals are merged. If there are additional optimized factors, fractal also integrates them. The goal of a fractal (g_f) can be optimized as follows:

$$g_f = F_f \oplus g_1 \oplus g_2 \oplus \ldots \oplus g_n \tag{7.2}$$

Equation 7.2 optimizes fractal f. Depending on the fractal objective, during optimization processes, the model can be minimized or maximized. However, goal coordination should be considered to optimize goals. For example, if the model includes fractal costs, the cost must be minimized. If the fractal bar decides to minimize the objective, the other objectives must also be minimized. Integration of numerical models can be repeated to any level of the fractal in the same way shown in Eq. 7.2.

Complex high-level fractal models can be constructed from a simple low-level fractal model by repeatedly fitting simple models. The adaptation at each level results are used not only to control the internal processes of the fractal but also to provide information for higher level fractals.

3.2 Mathematical Research Model for Cost Analysis

The variables, parameters, and indicators used in the introduced FRRP model are as follows:

Indicators:

i: Sub-fractals set ($i = 1, 2 \ldots, n$)
j: Resources set ($j = 1, 2, 3$)
j': Type of resource from the resources set ($j' = 1, 2 \ldots, m$)

Parameters and variables:

$hR_{ij}^{j'}$ is the cost of maintaining the i-th sub-fractal inventory for the j'-th type from the j-th resource per unit of time.
$cD_{ij}^{j'}$ is the cost of ordering the j'th type from the jth source of the ith sub-fractal per time unit.
$tc_{ij}^{j'}$ is the cost of transportation in the sub-fractal ith for the j'th type from the jth resource per time unit.
$tct_{jii'}^{j'}$ is the transportation cost of type j' from source j between sub-fractal i' and sub-fractal i'.
$cB_{ij}^{j'}$ is the delay cost of the j'th type from the jth source of the ith sub-fractal per time unit.
C_i is the cost expected of i sub-fractal per time unit.
C_f is the cost expected of fractal f with replenishment $q_{ij}^{j'}$ amount per time unit.
$x_{ij}^{j'}$ is the demand for the j'th type of the jth resource of the ith fractal sub in L period (probability function $= f\left(x_{ij}^{j'}\right)$).
L is the shipping time.
$S_{ij}^{j'}$ is the maximum inventory level for the j'th type of the jth source of the ith sub-fractal.
$s_{ij}^{j'}$ is the inventory level at the ith sub-fractal order point for the jth resource of the j' th type.
T_r is the ordering point.
T_p is the expected time between order points.
$qt_{jii'}^{j'}$ is the transported value of the j'th type from the jth source between the ith sub-fractal and the i'th sub-fractal.
$q_{ij}^{j'}$ is the amount of replenishment of the j'th type from the jth source of the ith sub-fractal.
$E(C_{\text{inv}i})$ is the expected cost of the ith sub-fractal inventory.
$E(C_{\text{BC}i})$ is the expected cost of delay of the ith sub-fractal.

C^f is the fractal overhead cost f includes shipping and ordering costs.
C_{qi} is the cost of sub-fractal i.
F_f is the numerical expression of additional factors for fractal f.
g_i is the goal of sub-fractal i.
g_f is the goal of fractal f.
$\oplus$ is the symbol for merging models.
caR_i is the capacity of the sub-fractal i m stock.
caD is the inventory capacity of fractal f.

Equation 7.3 shows the total cost of the fractal f. Equation 7.4 shows the cost of maintenance, delay, transportation, and order costs. The first part of Eq. 7.4 is the formula for the situation in which the demand ($x_{ij}^{j'}$) is smaller than the current inventory level. As shown in the second part of Eq. 7.4 if the demand is greater than the current inventory level, the delay cost should be considered in the numerical model. Equation 7.5 shows the expected time between order points. Also, Eq. 7.6 shows the expected time between order points.

$$C_f = \sum_{i=1}^{n} \sum_{j=1}^{3} \sum_{j'=1}^{m} \left[\frac{C(T_p)}{T_p} \right] + \sum_{i=1}^{n} C_i = \sum_{i=1}^{n} \sum_{j=1}^{3} \sum_{j'=1}^{m} \times \left[\frac{C(L) + C(T_p - L)}{T_p} \right] + \sum_{i=1}^{n} C_i \tag{7.3}$$

$$C(L) = \int_0^{s_{ij}^{j'}} \left[\left\{ \frac{hR_{ij}^{j'}}{2} * x_{ij}^{j'} + hR_{ij}^{j'} * \left(s_{ij}^{j'} - x_{ij}^{j'} \right) \right\} * L \right] * f\left(x_{ij}^{j'}\right) * dx_{ij}^{j'} + \int_{s_{ij}^{j'}}^{\infty} \left\{ \frac{hR_{ij}^{j'}}{2} * \frac{L * s_{ij}^{jj'^2}}{x_{ij}^{j'}} + \left(x_{ij}^{j'} - s_{ij}^{j'} - qt_{jii'}^{j'} \right) * cB_{ji}^{j'} \right\} * f\left(x_{ij}^{j'}\right) * dx_{ij}^{j'} + tc_{ij}^{j'} * q_{ij}^{j'} + tct_{jii'}^{j'}, * qt_{jii'}^{j'} + cD_{ij}^{j'} \tag{7.4}$$

$$C(T_p - L) = \frac{hR_{ij}^{j'}}{2} * \left(q_{ij}^{j'} - s_{ij}^{j'} \right) \Big\} * (T_p - L) + hR_{ij}^{j'} * s_{ij}^{j'} * (T_p - L) \tag{7.5}$$

$$T_p = \frac{\left(q_{ij}^{j'} - s_{ij}^{j'} \right) * L}{\left[\int_{-\infty}^{\infty} x_{ij}^{j'} * f\left(x_{ij}^{j'}\right) * dx_{ij}^{j'} \right]} + L \tag{7.6}$$

Here, $(x_{ij}^{j'} - s_{ij}^{j'})$ refers to the amount of delay. In this model, the transportation cost per unit from the source is considered and the cost of the order is considered in one unit of the order. Some constraints such as the service level of the company are ignored in this mathematical model due to the complexity of the formula. Without changing the model, the proposed cost model can be used in all fractal levels are used in the company set. Of course, the set of sub-fractals and resources depends on the fractal level.

The total costs related to the inventory management of the whole group can be calculated by summing the costs of all the sub-fractals in the supply chain. The results of the above analysis will be used to optimize the target of each fractal.

3.3 *Research Optimization Model*

Assume that the objective of each fractal is to minimize cost with respect to inventory management. Since the fractals themselves are similar, the target model of this fractal is the same in all fractal levels. The objective model with two decision variables is shown as follows:

$$g_f = \min C_f\left(q_{ij}^{j'}, qt_{jii'}^{j'}\right) \bigoplus g_1 \bigoplus g_2 \bigoplus \cdots \bigoplus g_n \tag{7.7}$$

Subject to:

$$\sum_{j=1}^{3} \sum_{j'=1}^{m} q_{ij}^{j'} \leq caR_i \quad \forall i \tag{7.8}$$

$$\sum_{i=1}^{n} \sum_{j=1}^{3} \sum_{j'=1}^{m} q_{ij}^{j'} \leq caD \tag{7.9}$$

$$\sum_{i=1}^{n} qt_{jii'}^{j'} \leq s_{i'j}^{j'} \quad \forall i, , j, j' \tag{7.10}$$

$$q_{ij}^{j'} \geq 0 \quad \forall i, , j, j' \text{ (integer value)} \tag{7.11}$$

$$qt_{jii'}^{j'} \geq 0 \quad \forall i, ,j,j' \text{ (integer value)} \tag{7.12}$$

Equation 7.7 is the objective function of the model. The main goal in the above model is to minimize fractal inventory costs. The presented objective function includes costs of backlog, inventory holding, ordering, shipping, and goods transfer. Equation 7.8 shows that the inventory level value for sub-fractal i must be smaller than the inventory capacity of sub-fractal i. Equation 7.9 shows the limit of fractal capacity f. Equation 7.10 means that the amount of transfer of goods from sub-fractal i' to all sub-fractals except i' must be smaller than the inventory level at the reorder point of sub-fractal i' ($s_{i'j}^{j'}$). $q_{ij}^{j'}$ in Eq. 7.11 shows the amount of replenishment of the j 'th type from the jth source of the ith sub-fractal. In addition, $qt_{jii'}^{j'}$ in Eq. 7.12 demonstrates the transported value of the j'th type from the jth source between the i-th sub-fractal and the i'th sub-fractal. Given s, S, and constraints, the inventory level value (q) can be obtained using the following equation:

$$q_{ij}^{j'} = \begin{cases} S_{ij}^{j'} - s_{ij}^{j'} + \dfrac{S_{ij}^{j'} - s_{ij}^{j'}}{T_r - T_s} * L & \text{if } \dfrac{S_{ij}^{j'} - s_{ij}^{j'}}{T_r - T_s} * L < S_{ij}^{j'} \\ S_{ij}^{j'} & \text{otherwise} \end{cases} \tag{7.13}$$

In Eq. 7.13, $q_{ij}^{j'}$ illustrates the amount of replenishment of the j'th type from the jth source of the ith sub-fractal. Transfer of resources between sub-fractals is done when $\left(\frac{S_{ij}^{j'} - s_{ij}^{j'}}{T_r - T_s} * L \geq S_{ij}^{j'}\right)$ and the amount of transport $\left(qt_{jii'}^{j'}\right)$ can be obtained using the following objective function:

$$\text{Minimize} \sum_{i=1}^{n} \sum_{j=1}^{3} \sum_{j'=1}^{m} qt_{jii'}^{j'} * tct_{jii'}^{j'} + \sum_{i=1}^{n} \sum_{j=1}^{3} \sum_{j'=1}^{m} \left(\frac{S_{ij}^{j'} - s_{ij}^{j'}}{T_r - T_s} * \left(L - s_{ij}^{j'} - qt_{jii'}^{j'}\right)\right) * cB_{ij}^{j'} \tag{7.14}$$

Subject to:

$$qt_{jii'}^{j'} \leq s_{ij}^{j'} \quad \forall i, i', j, j' \tag{7.15}$$

Equation 7.14 optimizes the fractal inventory cost f. Equation 7.15 shows the displacement limit of sub-fractals.

Minimize Z

$$
= \sum_{i=1}^{n}\sum_{j=1}^{3}\sum_{j'=1}^{m}\left[\frac{\int_{0}^{s_{ij}^{j'}}\left[\left\{\frac{hR_{ij}^{j'}}{2} * x_{ij}^{j'} + hR_{ij}^{j'} * \left(s_{ij}^{j'} - x_{ij}^{j'}\right)\right\} * L\right] * f\left(x_{ij}^{j'}\right) * dx_{ij}^{j'} + \int_{s_{ij}^{j'}}^{\infty}\left\{\frac{hR_{ij}^{j'}}{2} * \frac{L * s_{ij}^{jj'^2}}{x_{ij}^{j'}}\right.}{\frac{\left(q_{ij}^{j'} - s_{ij}^{j'}\right) * L}{\left[\int_{-\infty}^{\infty} x_{ij}^{j'} * f\left(x_{ij}^{j}\right) * dx_{ij}^{j}\right]} + L}\right.
$$

$$
+ \frac{\left.\left(x_{ij}^{j} - s_{ij}^{j'} - qt_{jii'}^{j'}\right) * cB_{ji}^{j'}\right\} * f\left(x_{ij}^{j'}\right) * dx_{ij}^{j'} + tc_{ij}^{j'} * q_{ij}^{j'} + tct_{jii'}^{j'} * qt_{jii'}^{j} + cD_{ij}^{j'}}{\frac{\left(q_{ij}^{j'} - s_{ij}^{j'}\right) * L}{\left[\int_{-\infty}^{\infty} x_{ij}^{j'} * f\left(x_{ij}^{j}\right) * dx_{ij}^{j}\right]} + L}
$$

$$
\left. + \frac{\left.\frac{hR_{ij}^{j'}}{2} * \left(q_{ij}^{j'} - s_{ij}^{j'}\right)\right\} * \left(\frac{\left(q_{ij}^{j'} - s_{ij}^{j'}\right) * L}{\left[\int_{-\infty}^{\infty} x_{ij}^{j'} * f\left(x_{ij}^{j}\right) * dx_{ij}^{j}\right]}\right) + hR_{ij}^{j'} * s_{ij}^{j'} * \left(\frac{\left(q_{ij}^{j'} - s_{ij}^{j'}\right) * L}{\left[\int_{-\infty}^{\infty} x_{ij}^{j'} * f\left(x_{ij}^{j}\right) * dx_{ij}^{j}\right]}\right)}{\frac{\left(q_{ij}^{j'} - s_{ij}^{j'}\right) * L}{\left[\int_{-\infty}^{\infty} x_{ij}^{j'} * f\left(x_{ij}^{j}\right) * dx_{ij}^{j}\right]} + L}\right]
$$

$$
+ \sum_{i=1}^{n}\sum_{j=1}^{3}\sum_{j'=1}^{m} qt_{jii'}^{j} * tct_{jii'}^{j} + \sum_{i=1}^{n}\sum_{j=1}^{3}\sum_{j'=1}^{m}\left(\frac{S_{ij}^{j} - s_{ij}^{j}}{T_r - T_s} * \left(L - s_{ij}^{j} - qt_{jii'}^{j}\right)\right) * cB_{ij}^{j'} \tag{7.16}
$$

3.4 The Final Research Model

The target model used in the research with two decision variables and problem constraints is shown as follows:

Subject to:

$$
\sum_{j=1}^{3}\sum_{j'=1}^{m} q_{ij}^{j'} \le caR_i \quad \forall i \tag{7.17}
$$

$$
\sum_{i=1}^{n}\sum_{j=1}^{3}\sum_{j'=1}^{m} q_{ij}^{j'} \le caD \tag{7.18}
$$

$$\sum_{i=1}^{n} qt_{jii'}^{j'} \leq s_{i'j}^{j'} \quad \forall i',,j,j' \tag{7.19}$$

$$q_{ij}^{j'} \geq 0 \quad \forall\, i,,j,j' \text{ (integer value)} \tag{7.20}$$

$$qt_{jii'}^{j'} \geq 0 \quad \forall\, i,,j,j' \text{ (integer value)} \tag{7.21}$$

$$qt_{jii'}^{j'} \leq s_{ij}^{j'} \quad \forall\, i,,j,j' \text{ (integer value)} \tag{7.22}$$

$$q_{jii'}^{j'} \leq S_{ij}^{j'} \quad \forall\, i,,j,j' \text{ (integer value)} \tag{7.23}$$

4 Research Implication

The studied company is one of the largest contracting companies in the field of construction in the country. The projects of this company are in different geographical areas, and the way of providing the resources needed for the projects plays an essential role in the timely completion of the project and the profitability of the project and the company. Based on this, different solutions are usually considered to provide the resources needed by the projects according to the number and size of the projects. Figure 7.5 shows the location of existing projects.

To better show the geographical location of the projects, it has been avoided to show the projects in one province. For example, the studied company in the southern part of Fars Province has seven large projects of dam construction, water canal construction, and ancillary facilities and building construction. Figure 7.6 shows the location of projects in the southern part of Fars Province (using Google Earth software).

Due to the extent of projects in the whole country, for better management of projects, the division has been done based on geography to follow up and provide resources and materials needed for projects, as well as quick and effective communication with employers. Figure 7.7 shows the geographical division of existing projects into regions.

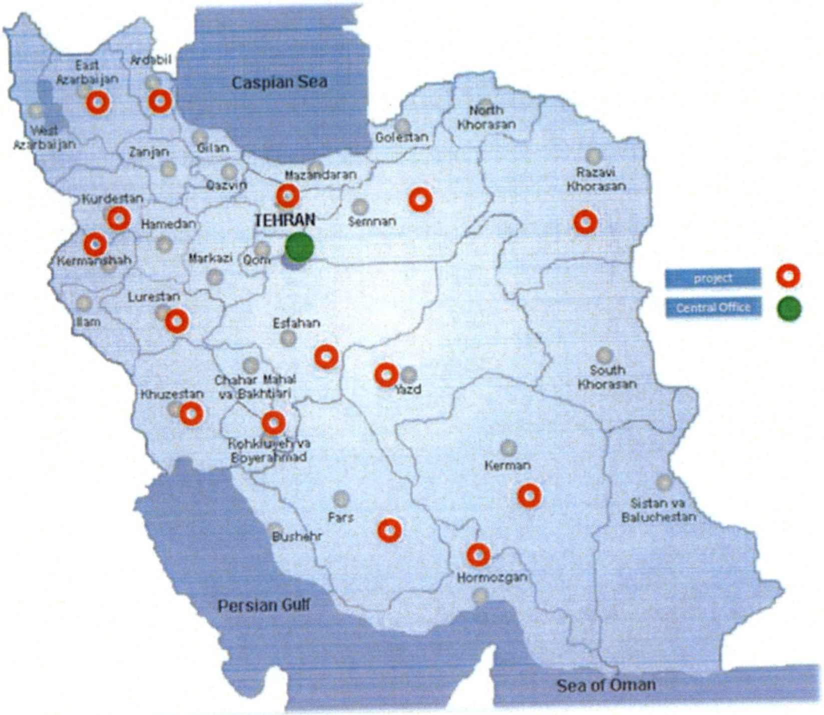

Fig. 7.5 Geographical location of existing projects

5 Research Results

To evaluate the proposed FRRP model, it was coded and run by GAMS software and the project information has been used as the input of this mathematical model. In Table 7.1 transferred values (from machines type 1), the project between and sub-fractals are shown.

Table 7.2 shows the transferred values (material type 1) between projects and sub-fractals.

Table 7.3 shows the transferred values (manpower type 1) between projects and sub-fractals.

The above tables show the results obtained from placing the information of the projects as the input of the proposed model, and finally, it's run by the GAMS software. These results illustrate that the introduced model is applicable as a case study and in real construction projects and industries. The results of the software are given in the appendix of this research for a better review. In this research, to model the research problem based on the conceptual model, the mathematical problem of resource and inventory management was developed based on a fractal system for purpose of minimizing the costs of transportation and movement of resources.

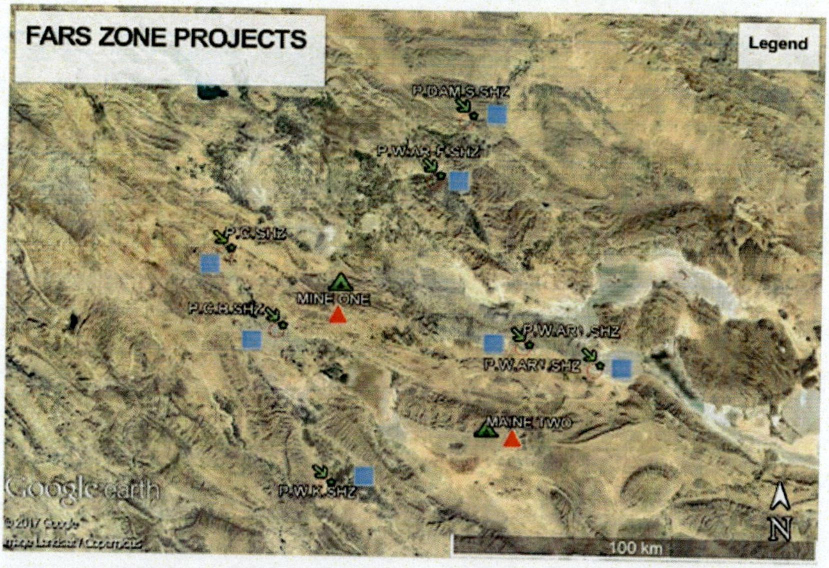

Fig. 7.6 Geographical location of existing projects in the southern part of Fars Province

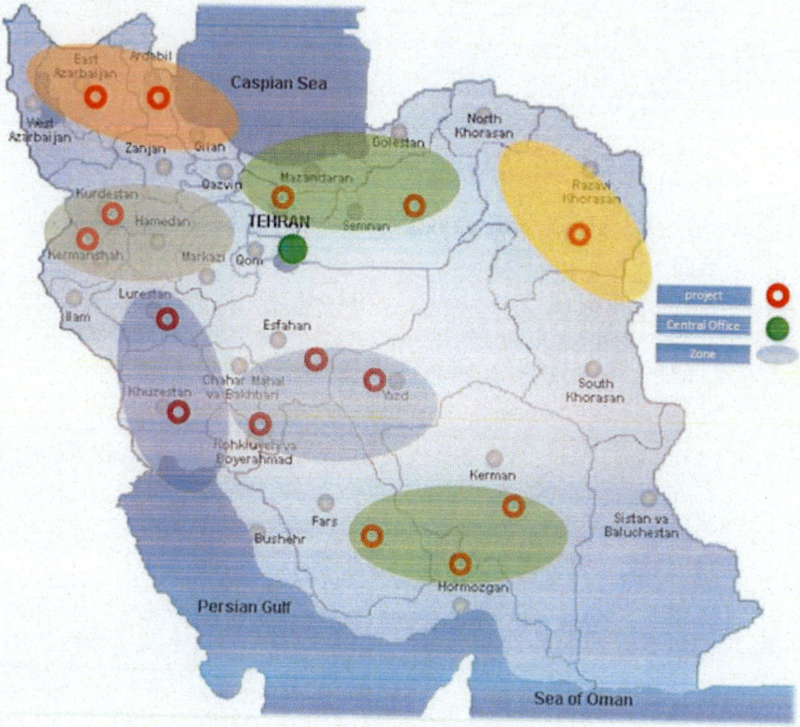

Fig. 7.7 Geographical division of existing projects into regions

Table 7.1 Transferred values of type 1 manpower

INDEX 1 = machine	Alamut	Shirkooh	Pataveh 1	Pataveh 2	Heart
1.alamut		40			
1.shirkooh	40				
1.pataveh2			30		40
1.herat				40	
2.pataveh1				50	80
2.herat	80	80	80		
3.alamut		100			
3.shirkooh	100			20	
3.pataveh1				100	20
3.pataveh2			20		100
3.herat	20	20	100		

Table 7.2 Transferred values of material type 1

INDEX 1 = material	Alamut	Shirkooh	Pataveh 1	Pataveh 2	Heart
1.alamut					40
1.shirkooh	40			40	
1.herat		40	40		
2.shirkooh	80				
2.pataveh1				80	80
2.pataveh2		80			
2.herat			80		
3.shirkooh	100		20	100	
3.pataveh1	20			20	20
3.pataveh2		100			100
3.herat		20	100		

Table 7.3 Transferred values of manpower type 1

INDEX 1 = human	Alamut	Shirkooh	Pataveh 1	Pataveh 2	Heart
1.shirkooh			40	40	
1.pataveh1	40	40			
1.pataveh2					40
2.alamut		80			
2.shirkooh	80		80		
2.pataveh1					80
2.herat				80	
3.alamut		20	20	20	100
3.shirkooh			100		20
3.pataveh1	100			100	
3.pataveh2		100			
3.herat	20				

6 Conclusion

In this research, the traditional concept of resource planning was changed. The headquarters managed the resource inventory of the regions, and the regions managed the resource inventory of the projects. Also, the possibility of supplying between group members in the same category to avoid delays and ensure the smooth flow of resources is considered. This research proposed a fractal resource requirement planning model in project-oriented construction and industrial companies by providing a fractal conceptual framework based on resource and inventory management to minimize resource costs and the smooth flow of materials between members to meet the existing demand.

To evaluate the proposed FRRP model, it was coded and run by GAMS software and the project information has been used as the input of this mathematical model. Due to the research question, after analyzing and results, it is specified that the proposed FRRP model could be used and implemented for resource management in project-oriented construction companies. According to the literature review that has been done in this research, it was found that little research has been done in relation to the use of fractal methods in resource management of companies and factories and proposing the FRRP model is for the first time in project-oriented construction and industrial companies, and this shows the novelty of this research.

The proposed FRRP model could be interested in academics, researchers, scholars, and anyone who is involved in projects. As future studies, the proposed FRRP model could be implemented in a real project. The implementation of this model could be useful in all kinds of projects and industries around the world to resource and inventory management to minimize the resource costs and the smooth flow of materials between projects to meet the existing demand.

References

1. Bider, I., Perjons, E., Elias, M., & Johannesson, P. (2017). A fractal enterprise model and its application for business development. *Software & Systems Modeling, 16*(3), 663–689. https://doi.org/10.1007/s10270-016-0554-9
2. Canavesio, M. M., & Martinez, E. (2007). Enterprise modeling of a project-oriented fractal company for SMEs networking. *Computers in industry, 58*(8-9), 794–813. https://doi.org/10.1016/j.compind.2007.02.005
3. Canavesio, M. M., Quaglia, C., & Martinez, E. (2016). Agent-based simulation of a project fractal company. In *XXII Congreso Argentino de Ciencias de la Computación (CACIC 2016).*
4. Chen, S. M., Chen, P. H., & Chang, L. M. (2012). Simulation and analytical techniques for construction resource planning and scheduling. *Automation in construction, 21*, 99–113. https://doi.org/10.1016/j.autcon.2011.05.018
5. Cheng, F., Li, H., Wang, Y. W., Skitmore, M., & Forsythe, P. (2013). Modeling resource management in the building design process by information constraint Petri nets. *Automation in Construction, 29*, 92–99. https://doi.org/10.1016/j.autcon.2012.08.005

6. Cheraghi, E., Zohrehvandi, S., Vanhoucke, M., & Tosarkani, B. M. (2022). A multiproject scheduling and resource management model in projects construction. *Engineering, Construction and Architectural Management.* https://doi.org/10.1108/ECAM-02-2021-0177
7. Karaa, F. A., & Nasr, A. Y. (1986). Resource management in construction. *Journal of Construction Engineering and Management, 112*(3), 346–357.
8. Karaulova, T., Poljantshikov, I., Shevtshenko, E., & Kramarenko, S. (2014). Fractal approach for manufacturing project management. *Mechanics, 20*(3), 352–359. https://doi.org/10.5755/j01.mech.20.3.6755
9. Liu, S. S., & Wang, C. J. (2007). Optimization model for resource assignment problems of linear construction projects. *Automation in Construction, 16*(4), 460–473. https://doi.org/10.1016/j.autcon.2006.08.004
10. Peralta, M. E., & Soltero, V. M. (2020). Analysis of fractal manufacturing systems framework towards industry 4.0. *Journal of Manufacturing Systems, 57*, 46–60. https://doi.org/10.1016/j.jmsy.2020.08.004
11. Rahmanniyay, F., & Yu, A. J. (2019). A multi-objective stochastic programming model for project-oriented human-resource management optimization. *International Journal of Management Science and Engineering Management, 14*(4), 231–239. https://doi.org/10.1080/17509653.2018.1534220
12. Sun, H., Liu, H., Tian, J., Guo, R., Xu, Q., Yao, L., et al. (2022). Modelling and optimizing resource management and environmental benefit of construction and demolition waste: A case study in China. *Buildings, 12*(9), 1361. https://doi.org/10.3390/buildings12091361
13. Warnecke, H. J. (1993). *The fractal company: A revolution in corporate culture*. Springer-Verlag.
14. Zohrehvandi, S. (2022). A project-scheduling and resource management heuristic algorithm in the construction of combined cycle power plant projects. *Computers, 11*(2), 23. https://doi.org/10.3390/computers11020023
15. Zohrehvandi, S., Vanhoucke, M., & Khalilzadeh, M. (2020). A project buffer and resource management model in energy sector; a case study in construction of a wind farm project. *International Journal of Energy Sector Management, 14*(6), 1123–1142. https://doi.org/10.1108/IJESM-10-2019-0025

Chapter 8
Smart Approaches by Online Monitoring in Transportation

Lucia Knapčíková, Annamária Behúnova, Jozef Husár, and Rebeka Tauberová

1 The State of the Art

From the point of view of freight transport management, the cargo must be in the right place at the right time, and it is true here that the better the management works, the more efficiently the transport works [1]. In operating freight transport, forwarding and forwarding companies play an important role. For a computerized forwarding system to be effective and efficient, it must contain programs that enable users to work efficiently for the needs of management. Processing must follow a logical flow of information related to the transportation of goods [2]. The processing must take place comprehensively in real time. The undercarriage of a rail vehicle is a device built between the tracks and the vehicle body or superstructure and serves to carry and guide the vehicle along the track. The chassis is usually connected to the main frame by pivot pins, firmly pressed into the chassis or the cabinet frame. It has some mobility compared to the lower side of the cabinet. Current trends in the construction of chassis are characterized by two directions that are interconnected. This involves increasing the axle weight to 25 t and increasing the maximum speed to 120 and 160 km/h. This is connected to the requirements for brake equipment, which must ensure safety even with new requirements [3–5].

L. Knapčíková (✉) · J. Husár · R. Tauberová
The Technical University of Košice, Faculty of Manufacturing Technologies with a seat in Prešov, Department of Industrial Engineering and Informatics, Prešov, Slovak Republic
e-mail: lucia.knapcikova@tuke.sk; jozef.husar@tuke.sk; rebeka.tauberova@tuke.sk

A. Behúnova
Technical University of Košice, Faculty of Mining, Ecology, Process Control and Geotechnologies, Institute of Earth Resources, Košice, Slovak republic
e-mail: annamaria.behunova@tuke.sk

L. Knapcikova, S. Zohrehvandi (eds.), *9th International Conference on Mobility, IoT and Smart Cities*, EAI/Springer Innovations in Communication and Computing,
https://doi.org/10.1007/978-3-031-28225-6_8

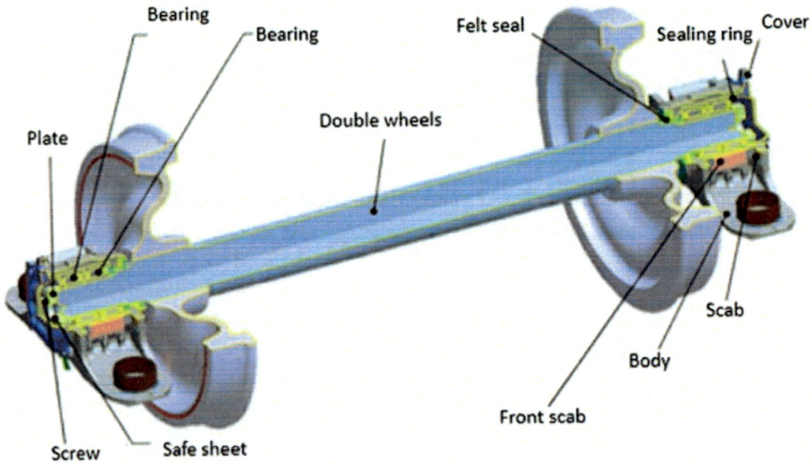

Fig. 8.1 Double wheels' construction [9]

The welded frame is checked for the quality of the welds and its required dimensions are verified. Subsequently, it undergoes surface treatments. A complete brake is mounted on the surface-treated frames [6]. At the end, the completely unfolded chassis frame with the brake is placed "lowered" onto the completed wheels with bearing chamber and axle bearings (Fig. 8.1) [7–9].

The chassis is a very important part of freight wagons. It is exposed to several acting forces. It is affected by the weight of the vehicle, payload, movement along the track, acceleration, braking, and various effects of devices attached to the wagon structure [7]. Due to the importance of the chassis for the reliability and safety of the rolling stock, demanding requirements are set for the chassis [9]. It is a matter of course that every manufactured chassis must pass a test – static loading of the chassis – before the sale (final clearance). The chassis loading device is used for this function [9].

1.1 Online Monitoring in Rail Transportation

The application of the online support system aims not only to increase safety but also to increase the cost of repairs of related components of the railway wagon [10]. Generating real-time information on the elements of the undercarriage of freight railcars, which carry a higher risk of damage during operation, would create ideal conditions for increasing the safety of their process [11]. The application of suitable technology can provide a solution for improving the quality of process of railway wagons. Online monitoring (Fig. 8.2) is of great importance in practical use, as it reduces the influence of the human factor acting in connection with technical inspections of

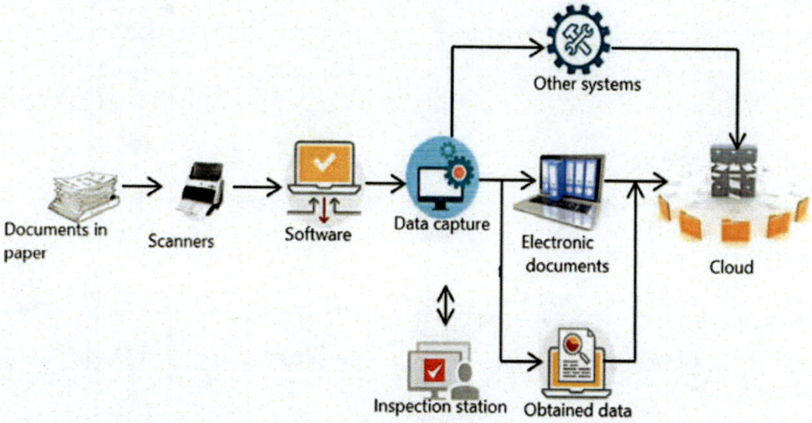

Fig. 8.2 Influence of the online monitoring process [6]

railway wagons [12]. Determining risk factors and their subsequent evaluation will provide information and thereby reduce rate of undesirable phenomena associated with the poor technical condition of freight rail wagons [8, 13]. Current production is still characterized by many outdated management tools, which describe further development [6].

Production processes are still recorded on paper – material tickets, work orders, or reports. In many plants, transcriptions from paper tickets to the enterprise resource planning (ERP) system are used, but they need to adequately cover the production processes. When ERPs started (Fig. 8.3) to take off in businesses two decades ago, they were expected to connect manufacturing with all other business systems, creating one compact whole [13, 14]. Expectations were fulfilled in many parts of the enterprises, but production was managed only in a few places. The problem is the collection of data from production. That is why it is very important to use information systems in the industry. There, paper tickets are collected and transcribed and put into the system [15, 16]. By introducing digital elements into processes and equipment, they see significant improvements in the use of equipment but also a noticeable reduction in the proportion of downtime [17].

As part of the wagon monitoring, it consists of applications for working with the wagon in the station with a subsequent link to the train inventory and its changes. In addition, the module includes supporting applications for weighing wagons and monitoring the condition of trains [16, 18]. Based on current assessments, digitization is used primarily in monitoring, data evaluation, and production planning. Enterprises are also recording improvements in the use of energy and media. According to current discussions, the challenge is also an opportunity to limit the possibility of human failure [19]. The number of reasons for introducing digitization is at least the same as the number of opportunities [18, 20].

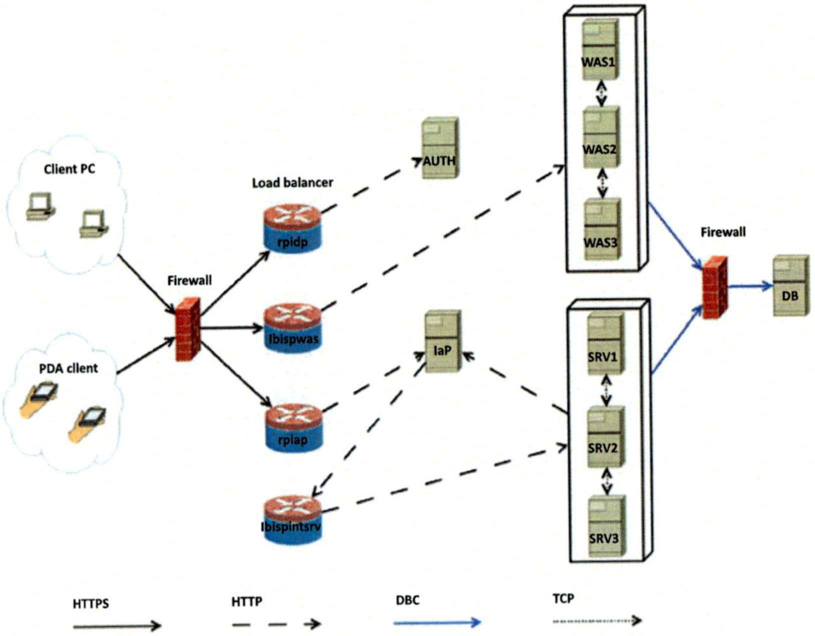

Fig. 8.3 Communication process in transportation [7]. Client PC = client workstations (with internet browser); PDA Client = industrial PDA; AUTH = authentication server; IaP = intranet server; WAS1, WAS2, WAS3 = application servers for interactive user access; SRV1, SRV2, SRV3 = application servers for noninteractive processing and process; DB = database server; Rpidp, lbispwas, rpiap, lbispintsrv, firewall = active elements of the network infrastructure (firewall, load balancer, reverse proxy) [7]

2 Work Methodology

An important task in the diagnostics of the technical condition of wagons is the pressure measurement of the rotating chassis [21]. It serves to measure and check various types of load-bearing and driving bogies. The loading device can act on the center of the chassis with a compressive force of 800 kN. With this device we can detect [18, 22] the following:

1. The size of the load force
2. Parallelism of axles
3. Functionality and tightness of the sensor
4. Functionality of brake cylinders
5. Chassis height
6. Pressure forces acting on the wheels

Basic parts of the measuring device [21, 22]:

The portal is a massive steel cross beam on which the print head and measuring unit for the control dimension X23 are placed. The portal is firmly connected to the portal supports by means of screws.

Load device – the power supply to the rotating chassis is mediated through the load device. The ring adapter, which is placed on it, ensures the pressing of the sliding insert into the chassis. The entire loading device is movable in the vertical direction.

Transport unit – moves along the rails and transports the rotating bogies to the measurement location.

Gauge leveler – a total of three gauge levelers are located on the movable side for adjusting the gauge and serve to level the wheel on the circular measuring plane when the chassis approaches.

Guide rails – are safety elements placed manually according to the wheel track setting. The guide rails prevent the swivel bogies from going to other gauges than the set gauge of the test bench.

The basic frame – consists of one self-contained steel welded structure, which, together with the travel unit, captures and distributes all the forces applied to the test bench. No forces are applied to the hall floor except for the self-weight of the test bench and the rotating chassis.

Portal support – supports the portal and superstructures mounted on it. Emergency switches and pneumatic maintenance units are lo the left and right supports.

Control unit X23 – measuring unit passing along the X axis and measuring the X23 dimension in the case of certain types of rotating chassis.

Stop – captures the kinetic energy of the rotating chassis and stops it during rolling inside the device. If the first two wheels pass through the stop, the reverse movement of the rotating chassis is no longer possible.

The bottom of the rail – captures the applied forces of the test bench. Shear force sensors, hunted under each wheel of the rotating chassis, ensure contact forces for each kneels during the loading process.

Moving unit – serves to readjust the gauge. Using the control panel, we can set the wheel track to 1435 mm and 1520 mm or 1668 mm.

Warning signal – the yellow warning signal lights up when a load operation is in progress. It signifies to workers that they are not allowed to enter the facility.

The lighting of the work area – two lamps on the pillars of the measuring bench guarantee sufficient lighting.

Compressed air and energy connections – a 230 V/400 V electrical source and a 10 bar pneumatic supply are located on the portal column [23–25].

Table 8.1 Load cycles

Cycle	Max. loading (kN)
1.	0–50
2.	50–100
3.	100–0
4.	0–100
5.	100–50

Authors own processing

2.1 Measurement and Load Process

A force of up to 100 KN acts on the chassis during loading. To ensure the correct position of various parts of the suspension (Table 8.1), it goes through several load cycles:

After the end of the first loading cycle, the first measurement $Z_7(1)$ and measurement Z6 take place, and after the end of the entire loading process, the following measurements are made [16, 22]:

1. Measuring the distance $Z_7(2)$ and Z_5
2. Measurement of functionality and tightness of the sensor
3. Distance measurement X_{23}

The load device contains a program in which the measurements of Z_5 and X_{23} are included. He records and evaluates the measured data with an average. The chassis that did not meet the measured values or the sensor function and pipe tightness check is moved for repair [23]. Next, the whole process is repeated [7].

After the load test, the following errors may occur [15, 24]:

1. Incorrect chassis height
2. Nonfunctioning load sensor
3. Inappropriate value of X_{23}
4. The height between the bearing housing and the chassis frame affects the overall height of the chassis.

Two parallel springs are mounted on each side of the bearing housing. The inner smaller spring starts to act only after a certain load on the outer spring. Only after the chassis frame drops (Fig. 8.4) [9, 22].

The distance between the bearing housing and the chassis frame is denoted Z_7.

Limit deviation for Z_7: $\pm$ 4 mm

Limit value for Z_7: 60 mm $\pm$ 4 mm

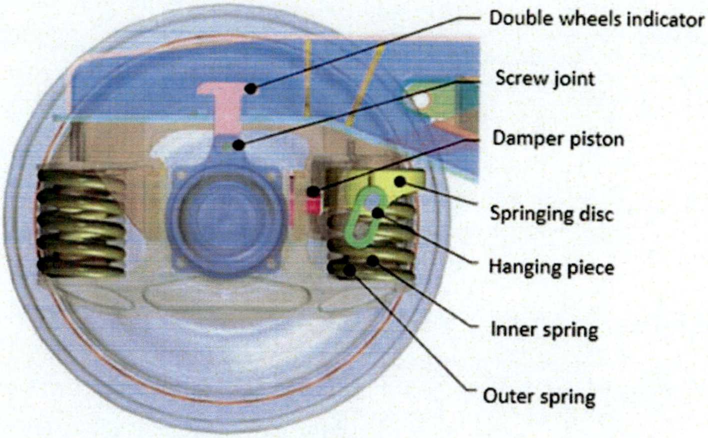

Fig. 8.4 Bearing box and chassis suspension [9]

3 Results and Discussion

The prerequisite for a correct chassis height analysis is the dimensions of the chassis frames manufactured within the prescribed tolerances and controlled by the specified measurement [16, 23, 24].

Z7 and Z5 distances were measured on a selected sample of 500 chassis. It is necessary to exclude measured values loaded with a gross error from the set of measured values.

1. Gross error calculation for $n = 500$ pcs.

The largest deviation from the average was measured on the first right wheel (Table 8.2).

From the ordered set of measured values, the following is calculated [6, 25]:

$$\text{Average value: } \overline{X} = \frac{\sum_{i=1}^{n} x_i}{n} = 62.042 \text{ mm} \tag{8.1}$$

and the standard deviation of the sample set:

$$S(x) = \sqrt{\frac{\sum_{i=1}^{n} (x_i - \overline{x})^2}{n-1}} = 1.509 \text{ mm} \tag{8.2}$$

Table 8.2 The test measurements Z_7

	1. test Z7				2. test Z7				Average value				Z_5
	R1	R2	L1	L2	R1	R2	L1	L2	R1	R2	L1	L2	
$\bar{X}$	62.0	61.8	62.7	62.8	59.2	59.2	59.8	59.81	60.66	60.52	61.3	61.31	−0.88
X_{max}	73.8	66.8	68.2	66.3	69.5	65.1	65.1	63.9	71.6	65.7	66.3	64.6	3.9
X_{min}	58	57	56.6	57.2	55.1	54.8	54	55.4	56.6	55.9	55.3	56.8	−4.6

Authors own processing

where:

x_i are the individual values of the file
n is the total number of values of the set

Calculation of standardized values:

$$H_1 = \frac{\overline{x} - x_{\min}}{S_{(x)}} = 2.677 \tag{8.3}$$

$$H_n = \frac{x_{\max} - \overline{x}}{S_{(x)}} = 7.787 \tag{8.4}$$

where:

$x_{\min}$ is the minimum value of the file
$x_{\max}$ is the maximum value of the selected file

The limit values of H for the selected probability P; the table limit value (H) is determined: $H = 4.607$ (from the table for the probability of 97.5% and $n = 500$).

Comparison of H_1 and H_n against H:

$$H_1 \langle H \tag{8.5}$$

$H_n \rangle H \Rightarrow$ the extreme value is loaded with gross error

The calculations show that the measurement is loaded with a gross error and will be excluded from the sample set (Table 8.3). The error calculation for $n = 499$ pcs.

From the modified set of measured values, it is calculated as indicated in Eqs. (8.1) and (8.2):

$$\text{Average value: } \overline{X} = \frac{\sum_{i=1}^{n} x_i}{n} = 62.018 \text{ mm} \tag{8.6}$$

and the standard deviation of the sample set:

$$S(x) = \sqrt{\frac{\sum_{i=1}^{n} (x_i - \overline{x})^2}{n - 1}} = 1.416 \text{ mm} \tag{8.7}$$

Similar to Eqs. (8.3) and (8.4), the standardized values are calculated:

$$H_1 = \frac{\overline{x} - x_{\min}}{S_{(x)}} = 2.837 \tag{8.8}$$

$$H_n = \frac{x_{\max} - \overline{x}}{S_{(x)}} = 2.882 \tag{8.9}$$

Comparison of standardized values against H ($H = 4.607$): $H_1 \langle H$ and $H_n \langle H$.

Table 8.3 Evaluation of measurements after exclusion of the file loaded with error

	1. test Z_7				2. test Z_7				Average value				Z_5
	R1	R2	L1	L2	R1	R2	L1	L2	R1	R2	L1	L2	
$\overline{X}$	62.0	61.8	62.7	62.8	59.2	59.2	59.8	59.8	60.6	60.5	61.2	61.3	−0.89
X_{max}	66.1	66.8	67.4	66.3	62.7	65.1	65.1	63.9	64	65.7	65.9	64.6	3.9
X_{min}	58	57	56.6	57.2	55.1	54.8	54	55.4	56.6	55.9	55.3	56.8	−4.6

Authors own processing

After completing all cycles of the loading process with a pressure force of 50 kN (simulation of a loaded wagon), we pressurize the sensor through the inlet pipe to a value of 5 ± 0.2 bar. After reaching this value, the pressure force on the chassis is gradually reduced to zero (empty wagon) [25, 26]. At the moment of zero compressive force, the program automatically records the amount of pressure in the load sensor. A positive result of this test of tightness and functionality of the load sensor is a minimal drop and stabilization of the pressure in the sensor [27, 28]. In case of a large drop in pressure, the tightness and connection of the pipeline are checked; in the extreme case, the entire load sensor is replaced.

4 Conclusion

Mobile network communication processes bring increased efficiency, productivity, and safety in the field of freight transport and logistics. New business opportunities are also emerging in the manufacturing and service sectors, in terms of automation and robotics, services for citizen mobility, solutions for more efficient logistics, or digitization of the entire transport system. This applies to large companies as well as small- and medium-sized enterprises, including start-ups. Digitalization of transport is primarily based on data management, as it is in every other sector [28]. From a business perspective, data can be considered a factor of production or raw material to be processed and refined to create added value. For this purpose, the free movement of data is essential. It is necessary to realize that information flows in transport arise between all subjects, i.e., between producers, suppliers, carriers, recipients, and consumers. At the same time, the information system cannot be considered as a separate isolated system, but the solution needs to be integrated into the entire transportation system. In the field of transport and logistics, information systems (IS) must offer the possibility of maximum automation of routine activities and increases in the efficiency of operations.

Acknowledgments This research was supported by the European Union's Horizon 2020 research and innovation program under grant agreement No.723274 and projects VEGA 1/0268/22 and KEGA 038TUKE-4/2022 granted by the Ministry of Education, Science, Research and Sport of the Slovak Republic.

References

1. Dong, X., Chen, B. M., Cai, G., Lin, H., & Lee, T. H. (2009). Development of a comprehensive software system for implementing cooperative control of multiple unmanned aerial vehicles. *Presented at the 2009 IEEE international conference on control and automation*.
2. Pajares, G. (2015). Overview and current status of remote sensing applications based on unmanned aerial vehicles (UAVs). *Photogrammetric Engineering & Remote Sensing, 81*, 281–330.

3. Shao, X., Liu, J. J., Gao, Z. M., & Wang, P. (2012). A study on intelligent onboard rail passenger service systems. In Y. Q. Ni & X. Ye (Eds.), *Proceedings of the 1st international workshop on high-speed and intercity railways* (pp. 69–82). Springer.
4. Cvitić, I., Peraković, D., & Kuljanić, T. M. (2017). Availability factors in delivery of information and communication resources to traffic system users. In *Smart solutions in today's transport. TST 2017* (pp. 28–41). Springer. https://doi.org/10.1007/978-3-319-66251-0_3
5. Peña Miñano, S., Kirkwood, L., Court, S., Farnsworth, M., Orlovs, I., Shehab, E., & Tinworth, N. (2017). A review of digital way finding technologies in the transportation industry. In *Advances in Transdisciplinary Engineering* (pp. 207–212). IOS Press BV.
6. Kanáliková, A., Behúnová, A., Mandičák, T., & Mesároš, P. (2019). New technologies for sustaining development in infrastructure, logistics and construction industry. *Acta Logistica, 6*(4), 171–177.
7. ZS CARGO. Description of the information system – Information system on support for ZSSK operation, CARGO (ISP). Accessed 09/15/2022. https://www.zscargo.sk/en
8. Mukhopadhyay, A., Hegde, R. R., & Thomas, A. S. (2019). Handover mechanisms in wireless heterogeneous telemedicine networks. *Presented at the 2019 3rd international conference on computing methodologies and communication (ICCMC).*
9. Husnjak, Š. A., Forenbacher, I., Peraković, D., & Cvitić, I. (2020). UAV forensics: DJI mavic air non-invasive data extraction and analysis. In *EAI MMS 2020*. Springer International Publishing, Cyberspace. https://doi.org/10.1007/978-3-030-67241-6
10. Wani, M. A., Bhat, F. A., Afzal, S., & Khan, A. I. (2020). *Advances in deep learning*. Springer.
11. Brtka, V., Jauševac, G., Jotanović, G., Stjepanović, A., & Stojičić, M. (2020). Identification of potentially hazardous traffic situations using deep learning. In *10th International conference on applied Internet and information technologies* (pp. 137–140). University of Novi Sad, technical faculty "Mihajlo Pupin".
12. Singh, P., & Manure, A. (2019). *Learn TensorFlow 2.0: Implement machine learning and deep learning models with Python*.
13. Kornaszewski, M., & Pniewski, R. (2019). The impact of new telematics solutions on the safety of railway traffic on the example of modern simulators railway traffic control devices. In J. Mikulski (Ed.), *Development of transport by telematics* (pp. 32–43). Springer.
14. Iatcheva, I. I., Andreev, A. D., Stancheva, R. D., & Lilyanova, I. T. (2016). Electromagnetic flow meter field distribution maximizing device sensitivity. *Materials Science Forum, 856*, 157–162.
15. Hensel, S., Marinov M. B., & Schmitt, M. (2020). System setup for synchronized visual-inertial localization and mapping. In *2020 XXIX international scientific conference electronics (ET), Sozopol, Bulgaria*.
16. Kumar, A., & Panda, S. P. (2019). A survey: How Python pitches in IT-world. In *2019 international conference on machine learning, big data, cloud and parallel computing (COMITCon)* (pp. 248–251). IEEE.
17. Boskov, I., Yetgin, H., Vucnik, M., Fortuna, C., Mohorcic, M. (2020, January) Time-to-provision evaluation of IoT devices using automated zero-touch provisioning. In *IEEE global communications conference GLOBECOM 2020 – Proc.*, vol. 2020.
18. Ahmed, K., Nafi, N. S., Blech, J. O., Gregory, M. A., & Schmidt, H. (2017). Software defined industry automation networks. In *2017 27th international telecommunication networks and applications conference (ITNAC)* (pp. 1–3). IEEE.
19. Demchenko, Y., Filiposka, S., Tuminauskas, R., Mishev, A., Baumann, K., Regvart, D., & Breach, T. (2015). Enabling automated network services provisioning for cloud based applications using zero touch provisioning. In *8th international conference on utility and cloud computing (UCC)* (pp. 458–464). IEEE.
20. Prajapati, A., Sakadasariya, A., & Patel, J. (2018). Software defined network: Future of networking. In *2018 2nd international conference on inventive systems and control (ICISC)* (pp. 1351–1354). IEEE.

21. Cvitić, I., Peraković, D., Periša, M., & Husnjak, S. (2019). An overview of distributed denial of service traffic detection approaches. *PROMET – Traffic & Transportation, 31*, 453–464. https://doi.org/10.7307/ptt.v31i4.3082
22. Grabara, J., et al. (2020). Impact of legal standards on logistics management in the context of sustainable development. *Acta Logistica, 7*(1), 31–37.
23. Hou, L., et al. (2021). Literature review of digital twins applications in construction workforce safety. *Applied Sciences, 11*(1), 1–21.
24. Mabkhot, M., Al-Ahmari, A., Salah, B., & Alkhalefah, H. (2018). Requirements of the smart factory system: A survey and perspective. *Machines, 6*, 23. https://doi.org/10.3390/machines6020023
25. Arnold, C., Kiel, D., & Voigt, K. I. (2020). Innovative business models for the industrial internet of things. In *26th international association for management of technology conference, IAMOT 2017* (pp. 1379–1396). https://doi.org/10.1007/s00501-017-0667-7
26. Illa, P. K., & Padhi, N. (2018). Practical guide to smart factory transition using IoT, big data and edge analytics. *IEEE Access, 6*, 55162–55170. https://doi.org/10.1109/ACCESS.2018.2872799
27. Herrmann, F. (2018). The smart factory and its risks. *Systems, 6*, 38. https://doi.org/10.3390/systems6040038
28. Cvitić, I., Peraković, D., Periša, M., & Botica, M. (2019). Novel approach for detection of IoT generated DDoS traffic. *Wireless Networks, 1*. https://doi.org/10.1007/s11276-019-02043-1

Chapter 9
Possibilities and Challenges for the Sustainable Development of Air Transport

Martina Koščáková, Peter Korba, Ingrid Sekelová, and Peter Koščák

1 Introduction

The emissions have been rising since 2014 despite innovations and more rules being put in place. The pollutants created by airplanes are a major environmental worry. The combustion of fuel concentrates CO2 directly into the atmosphere. The other gas that is created at higher altitudes during cruising is NOX. These two factors are in charge of 2–3% of the annual CO2 emissions worldwide [1].

The previous 10 years have seen a surge in the demand for air travel, with revenue-passenger-kilometers climbing by more than 5% per year. Despite the COVID-19 epidemic, recent projections indicate that demand for air travel will increase over the next 20 years at a rate between 2.4% and 4.1% a year [2, 3].

The relevance of sustainability in aviation has significantly increased during the past several years. On the regulatory front, governing organizations at the national and international levels have increased their push to cut carbon emissions. For instance, the EU joined the voluntary "Carbon Offsetting and Reduction Scheme for International Aviation" (CORSIA), which was started by the ICAO, in order to achieve the objective of attaining net zero emissions by 2050 [4]. On the consumer side, an increasing number of travelers are feeling compelled by the "flight shame" social phenomena to reduce their carbon footprint or use more ecologically friendly forms of transportation [5]. Additionally, airlines and their major business partners are cooperating to investigate and create strategies to lessen the environmental effect of the sector [6].

M. Koščáková (✉) · P. Korba · I. Sekelová · P. Koščák
Technical University of Košice, Faculty of Aeronautics, Košice, Slovakia
e-mail: martina.koscakova@tuke.sk; peter.korba@tuke.sk; ingrid.sekelova@tuke.sk; peter.koscak@tuke.sk

L. Knapcikova, S. Zohrehvandi (eds.), *9th International Conference on Mobility, IoT and Smart Cities*, EAI/Springer Innovations in Communication and Computing,
https://doi.org/10.1007/978-3-031-28225-6_9

Companies are choosing to become sustainable as a result of demand from the outside world, which is transforming the business climate. According to a 2017 research by Melkonyan, Gottschalk, and Kamath [7], changing business models now put sustainability as a top priority. Business activities have an impact on the environment and vice versa; hence, the impact of climate change should be considered while building a business strategy. The justifications for the change must be highly significant and pertinent because the procedure might be expensive at times. Globally, the current trend regarding climate change is becoming more and more well-known. Companies are under pressure to implement strategies that would help them position themselves as ecologically friendly [8].

The best ways to become more sustainable vary based on a variety of circumstances. Many companies are integrating sustainability into their everyday operations. There are various strategies to improve sustainability, and the choice of strategy is entirely up to the business. Some companies begin with modest initiatives, such as waste management, investing in environmentally friendly fuel sources, etc. Other companies select more expensive choices, to get long-term economic and environmental advantages. Future developments are more complex and advanced because innovation is moving at a faster rate than is ideal [8]. Companies all across the world are making several contributions to sustainability in an effort to lessen the adverse effects aviation has on the environment. Companies are modifying their organizational structures according to their industry and size, from biofuel to voluntary commercial offsetting programs to even fundamentally altering their core competencies [9].

2 Measures Toward Sustainability

The green aviation sector may quickly flourish thanks to innovation. Aviation has to take into account new technology as the key component to drive the transformation toward a green aviation industry development mode and accomplish the optimization of the industrial structure because it is a technology-intensive sector. Increased levels of scientific and technical innovation lead to improvements in the industry's overall safety, management effectiveness, cost management, and service standards, which boosts core competitiveness. Scientific and technical innovation must be prioritized in order to actualize the growth of the green aviation sector, and new technology research and promotion must get regulatory support. Therefore, a variety of initiatives are required to encourage innovation in the field of green aviation [10].

Therefore, a move toward new business strategies, cutting-edge technology, environmental legislation, and energy is required to achieve a green aviation sector mode. The use of aircrafts with high fuel-saving performance and low noise, elimination of aircrafts with high fuel consumption, bolstering of the development of advanced energy-saving technologies, improvement of fuel efficiency, reduction of environmental pollution, and attentiveness to climate change are just few ways airlines can work to lessen the environmental effects of aviation [11].

Table 9.1 Green innovations for the aviation industry

Category	Green solution
Energy carrier	Biofuels, fuel cells, hydrogen-powered aircraft, electric-powered aircraft, solar-powered aircraft, electric/hybrid ground vehicles, sustainable power for airports, greener helicopters, the return of the airship
Technology/design innovations	Geared turbofans (GTF), blended wing body aircraft, microwave dissipation of contrails, laminar flow wings, advanced air navigation, quiet aircraft, open-rotor engines, outboard horizontal stabilizers (OHS) configurations, air-to-air refueling of airliners, morphing aircraft
Materials	Advanced composites, metal composites, recycling
Operations	Wireless cabins, close formation flying, multimodal airports

Generally, there have been 25 main sustainable innovations, technologies, and practices identified by the Royal Aeronautical Society of the UK [12] that are seen as possible solutions for the aviation sector. We have sorted these innovations according to the general category under which they fall (see Table 9.1). In the next paragraphs, we focused more closely on a number of those solutions.

2.1 Organizational Initiatives

The first category aims to inadvertently green aviation. Organizational activities concentrate on reducing emissions by providing CO2 compensation schemes, assisting forestry efforts, or buying CO2 certificates. Advocates for this approach contend that organizational efforts are the simplest and least expensive to execute of all possible decarbonization strategies [13]. This claim is supported by the unavailability of aircraft technology that produces no emissions, as well as the expensive and limited availability of sustainable aviation fuels (SAF). There is a wide range of compensation systems available, with just a few airlines willingly purchasing CO2 certificates. Experts require the implementation of SAFs, due to the risks associated with organizational projects [14].

ICAO Carbon Offsetting and Reduction Scheme for International Aviation and the EU Emissions Trading System, multinational programs, seek to provide market-based frameworks for reducing GHG emissions while also spurring financial support for climate mitigation efforts. The ETS follows the cap-and-trade system, which limits yearly CO2 emissions from aviation by allocating emission allowances to airlines [15].

Moreover, stability in policy is essential for luring long-term investment. A strong regulatory framework for SAF is also emphasized. Stable policy environments are crucial, especially when it comes to fostering public-private partnerships and fostering an atmosphere that is conducive to investment. Economic issues and expenses are directly related to the function of government policy, indicating that government policy is required to secure capital to enable industrial scale-up [16].

Moreover, there is a goal of reducing travel volume and decreasing the number of travelers. Institutional factors, both formally through government-led regulations and informally through normative social obligations, constrain individual travel behaviors [17]. Aviation ticket taxes are one typical regulatory measure intended to restrict the amount of travel. These taxes use demand elasticity to indirectly lessen the environmental effect of aviation. Each leaving passenger must pay the tax, with the amount varying according to the distance to the final destination [18]. Sonnenschein and Smedy [19] discovered that if tax money weren't allocated for climate change mitigation and sustainable transportation initiatives, customers would be less inclined to pay [15].

Conway et al. [20] discovered that there were 95% fewer air passengers in 2020 than there were in 2019, based on their survey of over 1000 US adults. After the epidemic, respondents predicted changes to their future air travel, especially for business visits. 27% of business travelers anticipated a reduction in air travel, in part because of greater dependence on digital communication [20]. Moreover, "flight shame" has come to be recognized as a phenomenon linked to a decrease in a number of passengers, as well as to the social setting and how people interact with social norms. According to Wormbs and Soderberg's [21] examination of flight shame, people's decisions to cease flying are most influenced by their understanding of how flying affects the environment.

2.2 *Sustainable Aviation Fuels and Alternative Energy Sources*

One of the key measures toward sustainable aviation is the implementation of sustainable aviation fuels. SAF is different from conventional fuel in that it is produced using sustainable raw materials and renewable energy rather than fossil petroleum. The ability to cut CO2 emissions by up to 80% is the primary benefit of SAF over fossil fuels. SAFs may be blended with regular kerosene; therefore, they don't call for significant modifications to the infrastructure [14, 22].

There are essentially three generations of SAF. The first generation is fuels made from plants including canola, sugarcane, corn, soybeans, and palm oil. Despite lowering CO2 emissions, these fuels may interfere with food production, which limits the sustainability and scalability of biofuels. To overcome this conflict of interests, the second generation of SAF employs waste products such used cooking oil, municipal garbage, and water plants like algae. Nevertheless, second-generation SAF are not suited for the mass market because of land-use disputes and a lack of supply. There are attempts to overcome these challenges through the third-generation SAF, which, through the so-called "power-to-liquid" and "power-to-gas" processes, renewable energy produced by the use of wind, water, and solar power plants is utilized to collect water and CO2 from the atmosphere in order to manufacture synthetic kerosene using a challenging chemical process. Despite being completely carbon neutral and not threatening food production, production levels are still too low and costs are too high [14].

It is recognized that using hydrogen as fuel may propel an airplane, either directly through a hydrogen combustion engine or by converting it to energy in a fuel cell that powers an electric motor. Contrary to SAF, the use of hydrogen fuels necessitates modifications to the designs of aircraft and engines [23] as well as the creation of supporting infrastructure for the production, liquefaction, storage, and transportation of hydrogen. Similar to turboelectric aircraft, hybrid electric aircraft, or all-electric aircraft, new designs of aircraft power and propulsion systems will also be necessary. The deployment of electric aircraft would necessitate both the installation of essential charging infrastructure and an upgrade to the present power supply network [15, 24].

2.3 Technological Innovations

All modifications to an aircraft's hardware, whether they increase aerodynamic efficiency, reduce weight, or use an alternative, environmentally friendly power systems, fall under the category of technological innovation. Modernizing the fleet is the most popular strategy used by airlines to boost sustainability through technological advancement [14]. New engine technologies are being developed by US and European aircraft and engine manufacturers, together with a number of research institutions, to increase propulsion efficiency to lower fuel consumption and noise levels. How to strike a balance between the competing goals of lowering fuel usage and NOx emissions is a critical challenge in future engine design [22].

Aerodynamics is crucial for meeting environmental standards, advancing aircraft upgrades, and enhancing aircraft performance and economy. Adopting cutting-edge aerodynamic techniques to increase lift and decrease drag might reduce fuel consumption for airplanes already in use, lowering fuel costs, pollution emissions, and noise levels. Civil aviation authorities and environmental groups all over the world establish more stringent regulations and criteria for aircrafts when it comes to noise pollution. This is to be accomplished in three key areas: airframe noise management, engine, and nacelle noise reduction [24].

While the hardware changes only somewhat benefit the environment, alternative propulsion systems like battery-electric and hydrogen-powered systems may eventually prove to be a practical choice. Because of their reduced mechanical complexity and lack of (local) pollutants, electric engines provide several benefits, including notably fewer maintenance requirements. At the same time, aircrafts fueled by hydrogen are also under development. Airbus exhibited three hydrogen-powered, zero-emission concept planes in late 2020 that are anticipated to begin service in 2035 [14, 25].

Advanced materials like composite and Al-Li alloys are used to reduce aircraft weight, which in turn helps to reduce fuel consumption and engine emissions to some extent. These advancements are made in addition to propulsion system optimization and overall aerodynamic efficiency. Current green aircraft paint materials are particularly good at preventing corrosion and do not include chromate. They can efficiently minimize greenhouse gas emissions since they contain few volatile organic compounds [22].

Last but not least, the adoption of various IT solutions helps to enhance green manufacturing. The use of these tools in operation and process management facilitates information exchange and decision-making between suppliers, manufacturers, retailers, and customers. All aspects of green manufacturing may be replicated, shared, and tracked in a virtual environment made possible by IT technology, making it accessible to all stakeholders. Along with helping with aircraft production, IT also significantly contributes to carbon footprint reduction through paperless operations and smart cockpits [24].

Concerning energy intensity reduction, the ICAO Council adopted the CO2 emission standard for new aircraft in 2017, providing yet another incentive for technological advancement and application [26]. This standard regulates cruise fuel efficiency and, consequently, the CO2 emission for upcoming commercial aircraft and business jet deliveries. The ICAO decisions have now sparked a wide range of possibilities to reduce energy intensity, including changing airline operations, updating the fleet, introducing new aircraft, and increasing payload capacity [15].

The engine is the primary source of the aircraft's emissions. Fuel consumption and greenhouse gas emissions may be reduced right away by increasing engine performance and combustion efficiency [24].

Reduced fuel consumption may be also achieved by increasing cabin density or load factor. According to Morrell's [27] calculations, a short- and medium-haul aircraft might save 0.83% of its fuel for every 1% increase in seat capacity. However, after a pandemic, this choice may no longer be viewed as risk free for the general public's health [15].

One of the measures for the reduction of energy intensity is that based on their influence on the environment and fuel mix, individual flights made by various aircraft types would be taxed at various rates under the proposed pricing system. Higher takeoff and landing costs for less energy-efficient aircraft would act as an indirect incentive for airlines to upgrade or replace their fleet with more energy-efficient aircraft [8].

Greater efficiency in aircraft routing might, on a more general level, lead to fuel savings and so benefit green initiatives. Another option would be to provide passengers access to high-speed trains, which would reduce demand for short-haul flights (up to 400 km). By converting to electricity for all on-ground activities in the future, certain technologies may be able to reduce the amount of fuel used while on the ground and avoid the associated emissions [28].

2.4 Operational Improvements

One of the most important changes needed in operations is the elimination of ATM inefficiencies, which result in dog-leg routes, stacking at congested airports, waiting in line for a departure time while engines are operating, etc. Operational changes, which may be made to virtually the entire global fleet, may have a more immediate and significant effect than the introduction of new engines and aircraft [22].

One particular operational strategy that has been promoted is the employment of medium-range aircraft for long-distance flights with intermediate pauses. An airplane with a design range of 5000 km would use 29% less fuel to complete a 15,000 km journey in three hops than it would do the same distance in a single flight. This estimate was made using a straightforward parametric analysis. Air-to-air refueling has been proposed as a way to enable medium-range aircraft to be deployed on long-haul missions without the need for intermediate refueling stops. Moreover, it is widely recognized that close formation flying may be used to increase range or decrease fuel consumption. The planes may take off from various airports and fly in formation over great distances before separating to land at certain locations [22].

Boeing is working now with a number of airports, airlines, and other partners internationally to provide "tailored arrivals" systems that help minimize fuel consumption, lessen controller strain, and improve scheduling and passenger connections. Additional controller automation techniques are required to maximize personalized arrivals [22].

The flight management system has gone through numerous phases as a crucial technological advancement that may decrease fuel consumption and flight intervals, improve airspace operating capability, guarantee flight safety and quality, and lessen pilots' burden. The evolution of the flight plan into real-time dynamic management based on 4D track control has become one of the next-generation flight management system's defining traits. The primary trend now is performance-based navigation, data links between space, sky, and ground, and 4D leading technology based on optimum flight paths [24].

3 Barriers to Sustainable Aviation

Although there are international initiatives for the reduction of the environmental impact of air transport, the efficacy of both CORSIA, ETS, and other programs is in doubt. The combined worldwide contributions of the two programs implemented concurrently are unlikely to appreciably lower aircraft emissions [15]. Most compensation organizations primarily concentrate on GHGs released by the flight and ignore condensation trails, which are considerably more detrimental. Moreover, because it takes time for newly planted trees to absorb CO2 emissions, reforestation programs give a time-shifted impact. As a result, forestry operations only start to show their effects after around 20 years. Supporting the growth of SAF hence results in faster compensation than reforestation programs [14].

In relation to the suggested airport fees for energy intensity of the flights, the findings from Thelle and Mie la Cour's [29] study on European airport competitiveness were that in order to remain competitive, airports choose to reduce airport fees during difficult economic times. In the middle of the industry's recovery from the epidemic, instituting climate-related takeoff and landing fees in addition to the ICAO suggested landing costs might result in these airports losing customers to no-fee rivals [15].

3.1 Barriers to Implementation of Sustainable Energy Carriers

One of the most significant expense elements for the majority of airlines is fuel [30]. The most common reason cited for not commercializing a SAF manufacturing business is cost, and these costs are directly related to the ensuing problems with logistics and quality control. Given the present SAF production routes and petroleum oil prices, the price of SAF is not competitive, and the expense of SAF production and logistics is preventing new players from entering the market and preventing scale-up [16]. Fuel providers and airlines are unlikely to replace conventional fuels with more environmentally friendly fuel types until synthetic fuels are equally priced or even less than conventional fuel or are compelled by strong governmental legislation [30]. Given the low availability of public finance and the high predicted production cost and unit pricing, national SAF manufacturing may not be as commercially viable [15]. Other significant barriers are infrastructure interface-related logistical and quality control problems. According to experts, regulations are needed to handle SAF transportation and resolve quality control challenges related to infrastructure [16, 30].

Safety is another reason for concern. SAF has several dependability problems, especially in difficult circumstances. These fuels' safety has not been established. This conclusion implies that perceptions of safety require more study and continue to pose a threat to the growth of the SAF industry [16].

Lacking the necessary institutional structures in the form of legislation and worldwide or at least regional plans for the adoption and promotion of SAF in the aviation industry and elsewhere is another problem [30].

Environmental concerns associated with the location of a SAF manufacturing industry are considered a significant hurdle. Decisions on where to locate a biorefinery may also be influenced by environmental nongovernmental organizations and local community resistance or support. Most people enjoy the concept of SAF but are not keen on having a refinery in their neighborhood because of the transportation costs and conflict with agricultural activities [16]. Due to feedstock rivalry, there are worries that large-scale SAF production might have adverse externalities for industries dependent on forestry, including paper and pulp, heat and electricity, and road transportation. Additionally, the potential commercialization of bio-based SAF may be hampered by the newly proposed EU forest policy, which exhorts EU member states to prioritize the use of forests as a carbon sink [15].

Other issues related to the implementation of SAF are capital accessibility, biofuel accessibility, and forest control [16]. SAF can be carbon neutral and not in direct competition with other economic outputs depending on the kind. However, research is still in its early stages, and a lack of economies of scale prevents competitive pricing and speedy market availability. Despite how promising SAFs may seem, industry insiders and experts advise against focusing on only one solution and instead encourage research and funding in a variety of other fields [14] (Fig. 9.1).

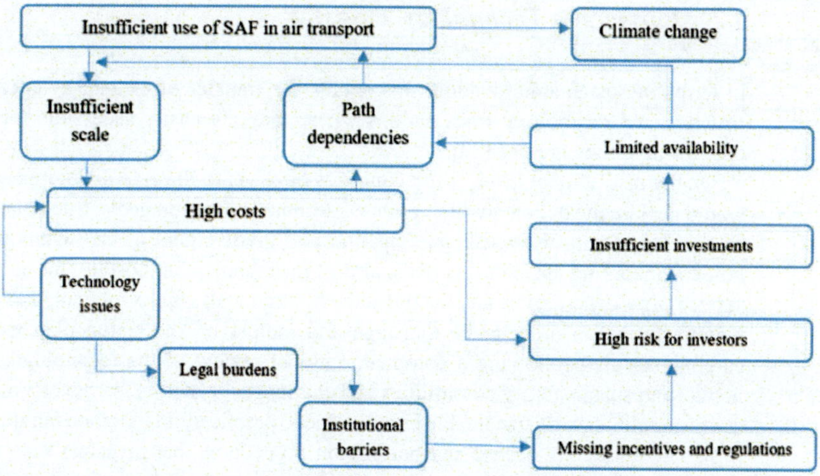

Fig. 9.1 Main barriers to the implementation of SAF in air transport. (Adapted from [30])

While hydrogen is a far more environmentally friendly alternative to traditional kerosene, there are several drawbacks to its use. First off, hydrogen cannot be produced on a large scale and requires a lot of energy. In addition, hydrogen needs to be kept at extremely low temperatures and high pressures, unlike kerosene. Last but not least, airlines need to invest in whole new aircraft fleets because their present fleets cannot run on hydrogen [14]. Other significant barriers to the use of hydrogen in aviation include high production costs, a lack of fuel infrastructure, and operating processes at airports. Public adoption of hydrogen-powered aircraft may also be hampered by a lack of general understanding regarding the safety of hydrogen propulsion technology. The few studies that have looked at the environmental effects of hydrogen propulsion have not taken into account the broader sustainability effects of the necessary infrastructure, such as dedicated green energy, storage, and transportation [15].

In the field of electric aircrafts, the employment of current batteries is restricted to smaller aircraft flying regional routes since they are currently too big and heavy to be deployed for long-haul flights. There aren't many instances of advanced project efforts because of this restriction and the early stages of development. Nonetheless, it is predicted that shortly, electric flying will be restricted to metropolitan regions [14]. However, the majority of current programs and projects are focused on technology, with little attention paid to the viewpoints of consumers. In their study on customers' desire to travel by electric aircraft, Han et al. [31] hypothesized that consumers' willingness to travel by electric aircraft may be influenced by factors such as green image, emotional attachments, attitudes, and moral standards.

3.2 Barriers to Technology Transfer

In terms of sustainable aviation, this entails the transfer of cleaner aviation infrastructure and technology from nations where they are easily accessible for use in others where they are not [32].

On the institutional level, many countries have yet to develop action plans which would declare the nation's will to pursue sustainable aviation goals. Such plans must be developed with the active participation and involvement of key aviation stakeholders, including airline executives, airport management authorities, air navigation service providers, airport and air traffic navigation staff, fuel suppliers, and aircraft manufacturers, in order to be thorough and inclusive. The action plan would be carefully created following a complete national review of the technological, economic, and sustainability possibilities and challenges related to the development and commercialization of sustainable aviation fuels, practices, and methods in the nation [32]. The numerous license and permission procedures that investors may have to deal with while attempting to bring in ESTs are one of the main concerns with technology transfer regulations in developing countries. Although these requirements must be met, navigating each bureaucratic agency's processes can be difficult and irritating. Eastern European nations are increasingly using the "one-stop shop method," which enables one organization to organize all necessary approvals [33]. So that investors won't need to deal with many licensing organizations. This method simplifies the technology transfer process, eliminates administrative bottlenecks, and ultimately lessens the stress and aggravation associated with registering technology transfer agreements. Governmental ministries, organizations, and departments involved in technology transfer must be coordinated, and countries must simplify their technology transfer procedures [32]. A sizable pool of officers and staff members that are trained and proficient in employing the newer and cleaner technology alternatives is required to support sustainable aviation. The mere importation or availability of new technology is insufficient; without the necessary human capability, such facilities may go unused or may not be used to their full potential. Governments must consequently prioritize determining the training requirements of regulators, airport management organizations, and airport employees and then work to supply them with sufficient and current information in key areas [32, 34].

On the legal level, the registration and regulatory approval processes for technology transfer are frequently stalled because of administrative inefficiencies and lack of transparency, which makes conducting business more difficult. Similar to this, the establishment of financial and policy incentive mechanisms that would encourage private sector involvement in technology creation and transfer has lagged in the development of technology transfer legislation. Moreover, the lack of effective intellectual property rights protection is a major impediment to the transfer of technology to many developing nations [10, 32].

3.3 *Barriers at the Airports*

The main obstacle to sustainable transformation at airports is the absence of recent legislative amendments and the growing need for new legislation since existing laws are frequently illogical, out-of-date, and slow to adapt to broader social, economic, and environmental developments. Second, there is a lack of training since many top management members do not have deeper knowledge about sustainability challenges [35]. Another issue hindering the development of sustainable airports is the high cost of new green technology. As there is a severe lack of official backing for a transition to green solutions, this issue is also tied to issues with finance [36].

According to a study by [37], several key elements are influencing the adoption of green technologies at airports. Technological aspects are among the most important ones. Without prior information, experience, or training about the notion, many airport operators find it challenging to comprehend the essence of green solutions. Another group comprises organizational aspects, as the awareness and support of top management are crucial for the adoption of new processes, and environmental measures are frequently perceived as being incompatible with current airport operations and procedures [36].

Another issue is that due to high construction costs and expensive certification fees, only a small number of airports focus on noise management and sustainable materials. Even though noise is considered to be the main environmental concern surrounding airports, measurements for its mitigation are among the least popular owing to the complexity of the issue. Additionally, there are safety concerns about airports becoming sustainable in terms of energy use, including glare, the reliability of new power systems, and the security of airside activities [38].

4 Discussion

In this part, we would like to present several recommendations for the development of sustainable aviation. As SAF were identified as the most viable mean of cutting CO2 emissions from the aviation sector, it is important to focus more attention on their implementation into use.

First of all, as private investments in SAF production are seen as highly risky, it is evident that private funding will not be sufficient to boost SAF wider production and governmental input is essential. Moreover, to support the development, national governments would be expected to step in for the support of new experimental SAF refineries, as well as help to make the components along the supply chain more affordable.

Secondly, it is important to implement such regulations that would encourage the use of SAF through several measures, such as various incentives, pricing, and subsidies to make them more competitive on the market and compensate for and reduce the potential risks related to their initial use. It is essential to reduce the enormous price difference between SAF and conventional fuels. One of the options would be to raise the fees associated with CO2 production during fossil fuels usage and at the same time impose tax reductions or exceptions for SAF usage. Moreover, it is advisable to introduce some kind of international blending quota that would require certain shares of SAF in aviation fuels. At the same time, it is important to pay attention to the whole SAF supply chain, lowering the input costs and reducing its carbon intensity to make sure SAF are indeed sustainable.

In relation to the promotion of sustainable technologies, it is important to speed up the transformation process of development patterns and focus on the development of new advanced commercial aircrafts that would comply with sustainability criteria, through the utilization of new materials, structures, design, and manufacturing technologies, to constantly limit emissions, noise, and fuel consumption. At the same time, the focus should be put on energy conservation projects and the sustainability of constructions.

On the operational level, higher efficiency can be achieved through the wider use of artificial intelligence (AI), which could help to optimize fuel consumption by evaluating historical flight data and effective flight routing. At the same time, AI could be used to optimize maintenance practices for aircraft components through data analysis and prediction of optimal maintenance periods. Among other recommendations, we can include support of research of possibilities of electric and hydrogen aircraft and encouragement of integrated transport system, due to which, short regional flights could be eliminated.

5 Conclusion

In this article, we have analyzed possible sustainable practices suitable for the development of sustainable air transport. In the first part, we have created an overview of sustainable practices that are currently in use or are being implemented. While there are 25 main sustainable innovations, technologies, and practices currently being developed for implementation in air transport, we have focused only on the most relevant of them. These measures can be sorted into four main categories – organizational practices, technological and operational innovations, and transfer toward sustainable energy carriers for aviation.

Through the analysis in the second part of the article, we have identified the three most relevant groups of barriers to sustainable development, which are related to technology transfer, green airports, and the shift to sustainable fuels. Among the key barriers identified were the high cost of new technologies, lack of regulations, lack of available innovations, focus on short-term goals, absence of clear benefits and goals, and guidelines without an actual action plan.

In the last part, we have focused on the recommendations for the sustainable development of air transport. We can conclude that most of the attention concerning environmental issues related to air travel is revolving around GHG emissions, partly ignoring other challenges. At the same time, the focus of green solutions in aviation is on the wider usage of SAF, which could bring the most significant cuts in emissions from the aviation sector.

For the future, it is important to focus on sustainable development in all the spheres of the aviation industry and aim for a coordinated international approach. The research should be extended beyond the ways for cutting CO2 emissions and focus on all the possibilities for making the aviation sector more sustainable.

References

1. *European Aviation Environmental Report 2019 – EASA.EUROPA.EU*. https://www.easa.europa.eu/eco/sites/default/files/2021-09/219473_EASA_EAER_2019_WEB_LOW-RES_190311.pdf
2. Lee, D. S., Fahey, D. W., Skowron, A., et al. (2021). The contribution of global aviation to anthropogenic climate forcing for 2000 to 2018. *Atmospheric Environment, 244*, 117834.
3. *Post-COVID-19 forecasts scenarios*. https://www.icao.int/sustainability/Pages/Post-Covid-Forecasts-Scenarios.aspx
4. *Aviation emissions: EU confirms its participation in the Corsia Voluntary Phase from 2021 and chooses more ambitious option to calculate its offsetting requirements*. https://www.consilium.europa.eu/en/press/press-releases/2020/06/25/aviation-emissions-eu-confirms-its-participation-in-the-corsia-voluntary-phase-from-2021-and-chooses-more-ambitious-option-to-calculate-its-offsetting-requirements/
5. *Why 'flight shame' is making people swap planes for trains*. https://www.bbc.com/future/article/20190909-why-flight-shame-is-making-people-swap-planes-for-trains
6. *A cleaner, quieter, smarter future for UK aviation*. https://www.sustainableaviation.co.uk/about-us/our-story/
7. Melkonyan, D. G., & V.K. V.P. (2017). Sustainability assessments and their implementation possibilities within the business models of companies. *Sustainable Production and Consumption, 12*, 1–15.
8. Jātniece E. *Drivers and barriers to sustainable aviation, example of Air Baltic*. http://hdl.handle.net/10062/68176
9. CAPA – Centre for Aviation Environmental Sustainability for Aviation. https://centreforaviation.com/analysis/airline-leader/environmental-sustainability-for-aviation-395173
10. Qiu, R., Hou, S., Chen, X., et al. (2021). Green aviation industry sustainable development towards an integrated support system. *Business Strategy and the Environment, 30*, 2441–2452.
11. Gegg, P., Budd, L., & Ison, S. (2015). Stakeholder views of the factors affecting the commercialization of aviation biofuels in Europe. *International Journal of Sustainable Transportation, 9*, 542–550.
12. Aerospace International. *The green Issue*.
13. *Enabling voluntary carbon offsetting*. https://www.iata.org/en/programs/environment/carbon-offset/
14. *Is sustainable aviation ready for take-off*. https://run.unl.pt/bitstream/10362/142626/1/2021-22_fall_44130_alexander-gerich.pdf
15. Lai, Y. Y., Christley, E., Kulanovic, A., et al. (2022). Analysing the opportunities and challenges for mitigating the climate impact of aviation: A narrative review. *Renewable and Sustainable Energy Reviews, 156*, 111972.

16. Smith, P. M., Gaffney, M. J., Shi, W., et al. (2017). Drivers and barriers to the adoption and diffusion of sustainable jet fuel (SJF) in the U.S. pacific northwest. *Journal of Air Transport Management, 58*, 113–124.
17. Jakobsen, M. *W. Richard Scott, institutions and organizations: Ideas, interests, and identities*. https://www.academia.edu/48613003/W_Richard_Scott_Institutions_and_Organizations_Ideas_Interests_and_Identities
18. *Tax rate per country – tax on air travel*. https://www.skatteverket.se/servicelankar/otherlanguages/inenglish/businessesandemployers/payingtaxesbusinesses/taxonairtravel/taxratepercountry.4.41f1c61d16193087d7f5472.html
19. Sonnenschein, J., & Smedby, N. (2018). Designing air ticket taxes for climate change mitigation: Insights from a Swedish valuation study. *Climate Policy, 19*, 651–663.
20. Conway, M. W., Salon, D., da Silva, D. C., et al. (2020). How will the COVID-19 pandemic affect the future of urban life? Early evidence from highly-educated respondents in the United States. *Urban Science, 4*, 50.
21. Wormbs, N., & Wolrath Söderberg, M. (2021). Knowledge, fear, and conscience: Reasons to stop flying because of climate change. *Urban Planning, 6*, 314–324.
22. Agarwal, R. K. (2009). Sustainable (green) aviation: Challenges and opportunities. *SAE International Journal of Aerospace, 2*, 1–20.
23. Baroutaji, T., Wilberforce, M. R., et al. (2019). Comprehensive investigation on hydrogen and fuel cell technology in the aviation and aerospace sectors. *Renewable and Sustainable Energy Reviews, 106*, 31–40.
24. Lin, Z.-M. (2013). Making aviation green. *Advances in Manufacturing, 1*, 42–49.
25. Marimuthu, S., Al-Rabeei, S., & Boha, H. A. (2022). Three-dimensional analysis of biomimetic aerofoil in transonic flow. *Biomimetics, 7*, 20. https://doi.org/10.3390/biomimetics7010020
26. International Civil Aviation Organization CO2 Standard for new aircraft, https://theicct.org/publication/international-civil-aviation-organization-co2-standard-for-new-aircraft/.
27. Morrell, P. (2009). The potential for European Aviation CO2 emissions reduction through the use of larger jet aircraft. *Journal of Air Transport Management, 15*, 151–157.
28. *IEA transport – Analysis*. https://www.iea.org/reports/transport.
29. Thelle, M. H., & la Sonne, M. (2018). Airport competition in Europe. *Journal of Air Transport Management, 67*, 232–240.
30. Scheelhaase, J., Maertens, S., & Grimme, W. (2019). Synthetic fuels in aviation – Current barriers and potential political measures. *Transportation Research Procedia, 43*, 21–30.
31. Han, H., Yu, J., & Kim, W. (2019). An electric airplane: Assessing the effect of travelers' perceived risk, attitude, and new product knowledge. *Journal of Air Transport Management, 78*, 33–42.
32. Olawuyi, D. S. *Sustainable aviation and the transfer of Environmentally Sound Technologies to Africa (Chapter 8) – Sustainable development, international aviation, and treaty implementation*. https://www.cambridge.org/core/books/sustainable-development-international-aviation-and-treaty-implementation/sustainable-aviation-and-the-transfer-of-environmentally-sound-technologies-to-africa/D1FAD5B462D2FBBD3ECEF00FA3990C3E
33. *Doing business 2011: Making a difference for entrepreneurs*. https://www.worldbank.org/en/news/press-release/2010/11/03/doing-business-2011-making-difference-entrepreneurs
34. Shah, K. J., Pan, S.-Y., Lee, I., et al. (2021). Green transportation for sustainability: Review of current barriers, strategies, and innovative technologies. *Journal of Cleaner Production, 326*, 129392.

35. Sullivan, C. *What challenges do airports face in their environmental sustainability commitment?* https://caterinasullivan.medium.com/what-challenges-do-airports-face-in-their-environmental-sustainability-commitment-4f1277210ac0
36. Korba, P., Koscakova, M., Fozo, L., et al. (2022). Current state and possible challenges in the development of green airports. *2022 New Trends in Civil Aviation (NTCA)*.
37. Harley, G., Timmis, A., & Budd, L. (2020). Factors affecting environmental practice adoption at small European airports: An investigation. *Transportation Research Part D: Transport and Environment, 88*, 102572.
38. Kis, F. *Green Airports Challenge for operators: How to support complex design decisions?* https://ratedpower.com/blog/green-airports/

Part III
Technological Innovations and IoT

Chapter 10
Opportunities for the Use of Biometric Technology in Air Transport

Juraj Horkay , Samer Al-Rabeei , Peter Korba , Michal Hovanec , and Volodymyr Tymofiiv

1 Introduction

Biometric identification is one of the most widespread elements in the development of modern airports. Airlines and airports around the world are actively working towards its introduction. Currently, the use of biometrics can be seen in several industries. For example, the use of gait recognition technology by security forces is becoming widespread, allowing individuals to be identified through their body shape and the way they move. As the aviation industry grows, so do security threats and the concerns are well founded. The pandemic has brought about the need for contactless passenger transfers through the airport. Contactless scanners provide faster passenger clearance through airport checkpoints. The airline industry is introducing more and more contactless self-service kiosks, booking, and payment options. Some companies have taken this a step further by introducing biometric facial recognition technology. Biometric facial recognition software can identify people by measuring dozens of facial features, allowing the face to be used to access facilities at the airport.

Nimra khan et al.'s study analyzes the Biometric Exit Program using biometrics at airports and identifies the challenges faced. A study of Dublin Airport's Entry-Exit Program is conducted, including facial recognition boarding gates. Challenges are

---

J. Horkay (✉) · S. Al-Rabeei · P. Korba · M. Hovanec · V. Tymofiiv
Technical University of Kosice, Faculty of Aeronautics, Department of Aviation Engineering, Kosice, Slovakia
e-mail: Juraj.horkay@tuke.sk; Samer.al-rabeei@tuke.sk; Peter.Korba@tuke.sk; Michal.hovanec@tuke.sk; Volodymyr.tymofiiv@tuke.sk

L. Knapcikova, S. Zohrehvandi (eds.), *9th International Conference on Mobility, IoT and Smart Cities*, EAI/Springer Innovations in Communication and Computing,
https://doi.org/10.1007/978-3-031-28225-6_10

identified based on the results of pilot tests at Dublin Airport and other U.S. airports. Dublin Airport and other US airports participated in pilot tests [1].

This technology has recently developed and advanced at an incredible rate. Next-generation biometrics has had a particularly large impact on the commercial aviation industry, with many airlines and airports around the world adopting biometric facial identification.

In conjunction with the Metropolitan Washington Airports Authority, Emirates Airlines began using biometric facial recognition technology to board passengers on its Dubai-bound flights from Washington Dulles International Airport this summer [2].

There has been a very little ethnographic examination of biometric identification (or IDentification) in the past few years. According to this argument, biometric technologies are promoted as a way to identify individuals objectively and incontestably, but in actuality, they are rooted in specific social contexts, fraught with ambiguity and uncertainty, and dependent on human interpretation. As a result, anthropology is very interested in and concerned about them [3].

Biometric identity systems can benefit from AI through face detection; however, this can be challenging due to the problems of continuous learning, decision-making, and security [4].

Unique, universal, and relatively unchanging biometric identification methods include fingerprints, iris scans, voice recognition, and facial features. Due to concerns about privacy and the potential for law enforcement to use biometric data, biometrics have not been widely implemented in developed nations [5] [6].

In Negri et al., it was looked into if airport visitors would use biometric technologies [7].

Research demonstrates that the use of biometrics and other essential technology improves the efficiency of passenger processing operations [8, 9].

This paper highlights the use of artificial intelligence to identify people in order to check-in passengers faster at airports, enhance airport security, and make air travel more enjoyable by reducing the time associated with screening and identifying passengers during check-in, passport control, boarding, and baggage check-in. The thesis is divided into the following parts: biometric methods of identification, biometric facial recognition (facial biometrics), use of biometrics in the present, facial biometrics in the world, legislative limitations and protection of personal data, and proposal for a methodology of application of biometrics in the airports of the Slovak Republic.

2 Biometrics Technology and the History of Its Development in the World

The term "biometrics" is derived from the Greek words bio (life) and metria (measurement). Interestingly, the term biometrics was not used to describe these technologies until the 1980s. The first written reference found with the term "biometrics" was in 1981 in an article in The New York Times [10]. Biometrics is an automated way of recognizing people based on physiological or behavioral characteristics. The term biometrics can also be understood as the science of identifying or verifying the identity of persons based on physiological and behavioral characteristics.

The most common way of recognizing persons using biometrics is by analyzing a body part such as face, fingerprints, or iris of the eyes. There is also ongoing research in the area of vascular pattern recognition, hand geometry, DNA and even body odor. Handwriting, voice, and the way a person walks are also considered biometrics. Biometrics is the process by which a person's unique physical and other characteristics are detected and recorded by an electronic device or system as a means of confirming identity. The term "biometrics" refers to the statistical analysis of biological observations and phenomena [11]. Because biometric identifiers are unique to individuals, they are more reliable in verifying identity than knowledge-based methods such as ID cards and passwords.

Biometric identifiers are often referred to as "physiological" or "behavioral" identifiers.

Physiological biometric identifiers are related to the physics of the person and include the following:

- Fingerprint recognition
- Hand geometry
- Smell/odor
- Iris scans
- DNA
- Facial recognition

Behavioral identifiers are related to a person's behavior pattern and include the following:

- Keystroke dynamics
- Gait analysis
- Voice recognition
- Mouse usage characteristics
- Signature analysis

Biometric technologies are systems or applications that are designed to use biometric data derived from biometric identifiers. A biometric system is an automated process that:

- Collects or captures biometric data through a biometric identification device, such as a fingerprint or palm vein pattern scanner, or a camera to collect facial and iris images
- Compares the scanned data from those captured for comparison
- Compares the submitted sample with the templates
- Determines or verifies that the identity of the biometric data holder is authentic

Biometric technologies therefore consist of both hardware and software. A biometric identification device is hardware that collects, reads, and compares biometric data. Biometric data are sample taken from an individual that is unique to that individual's own person. The software embedded in biometric technology contains a biometric module that processes the collected biometric data. The software typically works in tandem with the hardware to operate the biometric data capture, data extraction, and matching process, including data matching.

Biometric technologies can be further classified according to the type of biometric data used in the system. The technologies are usually used either to identify persons and their characteristics with a database, such as criminal records, or to authenticate the identity of persons.

While early instances of biometrics date back to the Babylonian Empire, the first biometric identification system was not developed until the 1800s. Parisian Alphonse Bertillon created a method of classifying and comparing criminals based on their body measurements. Although Bertillon's system was imperfect, it was the catalyst for the use of physical characteristics as a means to verify identity. In the 1980s, when fingerprints began to be used as a way to identify criminals and to sign contracts, it was common knowledge that people had unique fingerprints that were a symbol of their identity. Although it is not 100% certain who first started using fingerprints for identification, we do know that Edward Henry developed the fingerprint standard of the Henry Classification System. This system was the first to be used for identification and was dependent on a unique fingerprint image. Law enforcement quickly discarded Bertillon's methods and began using Henry's classification system as the standard for criminal identity authentication. This triggered 100 years of research into other unique biological factors that could be used as a method of identification.

Biometric methods of recognition have been applied by mankind throughout its history. In fact, we most often recognize famous people precisely with their help – by face, voice, or gait.

Since the nineteenth century, biometric technologies, especially fingerprints, have been used in criminalistics, and since the end of the last century, as technology has developed, it has been possible to formalize algorithms for recognizing people by their appearance or behavior and to use automated systems to do so.

Biometric technology is currently experiencing a period of rapid development. This growth is in many ways linked to the decisions of leading governments to apply it to passport and visa documents, which have directed large financial and material resources into this area. There is also a huge public interest in these technologies.

The word "biometrics" is often found in various news reports on television, newspapers, and radio. Unfortunately, the people who use this concept do not always know exactly what they are talking about. This article is an attempt to clarify the basics of biometric technology, talk about how they work, where they can, and where they cannot be applied.

Biometrics solves the problems of authentication and identification... In the first case, the challenge is to ensure that the biometric characteristic obtained matches the one previously taken. Authentication (or 1:1 matching) is used to verify that the subject is who they say they are. A decision is made based on the degree of similarity of the characteristics.

Identification (or 1 to N comparison) addresses the issue of finding the most appropriate biometric characteristics from those previously used. In the simplest case, it is a sequential implementation of comparing the obtained characteristics with all available ones. In this case, the result is the most similar previously adopted characteristic (identification will not be performed at all) if the degree of similarity turns out to be lower than the degree indicated by all comparisons.

Let us see how biometric technologies work by using fingerprint recognition as an example. For recognition, it is necessary to obtain (with the help of special readers) an image of the papillary pattern of one or more fingers. Next, this image is processed, and its characteristic features, such as line branching, line ending, or line intersection, are found in the processing process. For each object, in addition to its type, the relative position and other parameters are also remembered, for example, for the end point – the direction of the line. The combination of these properties and their characteristics forms the biometric characteristics template.

When identification is used to compare the received template with previously obtained ones. At a certain level of matching, a conclusion is made about the identity of the templates and hence verification or identification of the submitted finger is made.

Recognition occurs in the same way for others. Naturally, other traits of characteristics are used, for example for the face – that is, the location and relative size of the nose, cheekbones, etc. In addition, because photographs can vary in size, it is necessary to change the scale to compare them, with the distance between the pupils used as the "scale."

The performance mark of biometric technology, in addition to cost and usability indicators, is based on the use of two probabilistic parameters – the false rejection rate (FRR) and the false acceptance rate (FAR). A wrong rejection error occurs when the system does not recognize a biometric function that matches the template it contains, and a wrong omission error occurs when the system incorrectly associates an offered function with a template that does not actually match it. As can be seen, the false skip error is a more dangerous safety in this respect, and the false reject error leads to a reduction in the usability of the system, which sometimes does not recognize the human the first time.

2.1 Classification of Biometric Technology

The advantages of biometric identifiers based on a person's unique biological, physiological characteristics that uniquely confirm identity have led to the intensive development of corresponding means. Biometric identifiers use static methods based on a person's physiological characteristics, i.e., the unique characteristics given to them from birth (finger papillary line drawings, iris, retinal capillaries, thermal

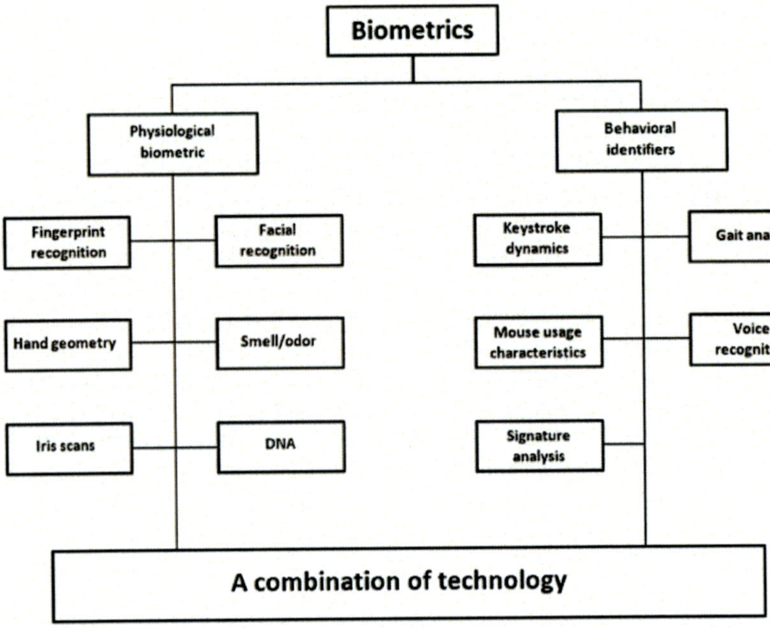

Fig. 10.1 A figure caption is always placed below the illustration. Short captions are centered, while long ones are justified. The macro button chooses the correct format automatically

imaging of the face, hand geometry, DNA) and dynamic methods (handwriting and signature dynamics, voice and speech characteristics, keyboard rhythm). It is proposed to use such unique static methods as identification by the underbite layer of the skin, by the volume of the fingers indicated on the scan, the shape of the ear, body odor, and dynamic methods – identification by the movement of the lips while playing a code word, by the dynamics of turning the key in the door lock, etc. A classification of modern biometric identification tools is given in Fig. 10.1.

Biometric identifiers only work well if the operator can check two things: first, that biometric data were received from a specific person during the check, and second, that these data match the sample stored in the card file. Biometric characteristics are unique identifiers, but the question of their reliable storage and protection from interception is still open.

Physiological biometrics is based on the classification of a person according to data collected as part of the human body, such as their fingerprints, face, or iris. Physiological characteristics are relatively stable. Face, fingerprints, silhouette of the hand, iris pattern, blood vessel, retinal pattern, or DNA fingerprints are biometrics that are essentially permanent and do not change over time. Behavioral data, on the other hand, are more susceptible to change depending on factors such as aging, injury, or even mood [12]. There are many techniques based on the seven factors shown in Fig. 10.1.

2.2 Division of Biometrics Technology Based on Physiology

Visual

- *Iris recognition*. The use of elements is found in the iris to identify an individual.
- *Retinal recognition*. The use of vein patterns are in the back of the eye to achieve recognition.
- *Facial recognition*. The analysis of facial features or patterns to authenticate or recognize the identity of an individual. Most facial recognition systems use custom shapes or local feature analysis.
- *Fingerprint recognition*. The use of ridges and valleys (markers) is found on the surface tips of a human finger to identify an individual.
- *Identifying an individual using the shape of the ear.*
- *Recognition of veins*. This is a type of biometrics that can be used to identify individuals based on vein patterns on the human finger or palm.

Visual-Spatial

- Finger geometry recognition. The use of 3D finger geometry to determine identity.
- Hand geometry recognition. Using hand geometric features, such as finger lengths and hand widths, to identify an individual.

Chemical

- Identifying an individual by analyzing segments from DNA.

Behavioral

- Gait. Using an individual's walking style or gait to determine identity.
- Recognition of writing. Use of a person's unique characteristics as handwriting.

Olfactory

- Smell/odor. Use of an individual's smell to determine identity.

Auditory

- Verification/authentication of voice.

3 The Use of Biometrics Today

Over the next 10 years, facial recognition is likely to transform airports more than any other technology. Airports in Panama, Dubai, Australia, the USA, and other countries have already started with facial recognition. It is only a matter of time before virtually every international airport adopts the technology. Facial recognition offers a wide range of security benefits that can improve various aspects of air travel. Facial recognition can also be used to protect passengers and make air travel safer than ever before.

3.1 Smarter Surveillance at Airport Entrances

There is no way to remember the name and face of every individual who is wanted, who is on a wanted list, or who is on a terrorist watch list. This is especially true if the criminal has changed appearance or is wearing a disguise. However, facial recognition can automatically identify when a person of interest enters the airport, and the system will not be fooled by a change in appearance. Facial recognition tracking works by capturing images of individuals and then using an algorithm to select the best image to compare against a database of photographs. If there is a match, the system immediately alerts security personnel. The system does not filter by age, gender, race, or country of origin [1]. And it is rarely fooled by a change in appearance. It has been shown to work well over long distances and is ideal for airport surveillance.

3.2 Intrinsic Threats at the Airport

One of the growing concerns in aviation is internal threats. These threats can range from theft of passenger baggage to threats that pose risks to public safety. Facial recognition can help by ensuring that employees only have access at the airport in areas where they are authorized to be and not where they should not be [13]. Facial recognition can instantly recognize employees and alert airport security if they are in areas where they are not allowed access.

3.3 Mobile Identification

Airport police sometimes stop cars as part of a random or scheduled security check. In these cases, airport police may use mobile facial recognition applications to instantly identify people in cars. By simply photographing a person from a safe distance, an officer can instantly determine if that person is a wanted criminal or on a terrorist watch list [25]. Airport police can also use mobile apps to instantly identify suspects in virtually any part of the airport.

3.4 Baggage Theft

Baggage theft is unfortunately a common occurrence at the airport. For example, according to available statistics, between 2010 and 2014 airlines processed 30,621 complaints regarding missing luggage. It is a problem especially at certain airports. Although it is currently not yet possible to completely prevent baggage theft, facial recognition can help to prevent theft by, for example, identifying already known persons in the baggage claim area who have already been caught in a similar incident [13].

4 Proposal for the Implementation of Facial Biometrics at the M. R. Štefánik Bratislava Airport

The aim of this thesis is to propose a suitable solution for the identification of persons using biometrics and artificial intelligence at the M. R. Štefánik Bratislava International Airport. The main reason for implementing this solution is to simplify and speed up the passenger handling process such as check-in, security check, passport control, baggage handling, and access to restricted areas at the airport. The new passenger identification technology will contribute to a more enjoyable overall air travel experience and enhance security at the airport. Based on the available applications used in foreign airports to date and the similarities between airports, it is appropriate to consider the introduction of facial biometrics for faster passenger processing and reduced dwell time in critical areas, as it is shown in the Fig. 10.2.

Online check-in has become the standard for many airlines in today's digital age. By using an online application or website, passengers can check-in for scheduled flights from the comfort of their home without standing in a long queue at the airport. In the case of online check-in, the use of facial biometrics is recommended as an ideal solution. When entering personal data as part of online check-in, the passenger takes a picture of his or her face using a mobile phone, which is used to create a facial signature or so-called Face ID. The photo, as well as the personal data associated with that photo, is then encrypted and sent to the CBP's Traveler Authentication Service system, which runs on a cloud server [15]. This server is shared between the airport and the airlines. Based on this information, it is evident that by linking and sharing the initial passenger information, M. R. Štefánik Bratislava Airport will receive information about arriving passengers in advance, which will allow the airport to eliminate potential threats and take action if a person from the wanted

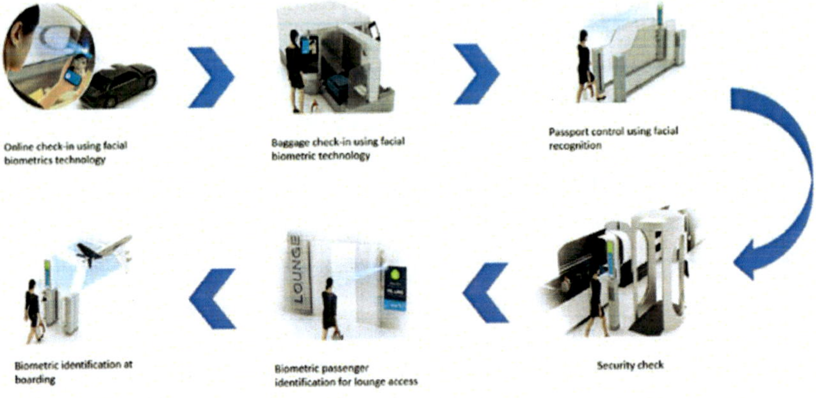

Fig. 10.2 Use of biometrics at the airport

person's database is among the passengers. This information is also applicable to other cases, such as various travel and medical restrictions.If this solution is implemented, the airport will have the advantage of a sufficient time margin to take the necessary measures of various natures, such as the readiness of security forces in the case of a wanted person (notification of law enforcement authorities) and the readiness of assistance services in the case of handicapped or physically disabled persons. On arrival at the airport, passengers will have the opportunity to use a self-service kiosk for checking in their luggage, both when boarding the aircraft and when entering the airport lounge. The procedure is described in more detail in the following chapters.

If the passenger has used online check-in for his/her flight, there is no need to wait in line again in the standard way. The passenger can proceed to automatic baggage check-in. After proceeding to the self-service baggage check-in belt, the passenger's face is captured by a camera. If the passenger has used the facial recognition service during online check-in and has created a Face ID, the passenger authentication service system compares the actual photo with the data stored in the cloud server and identifies the passenger. There is therefore no need for additional verification of the passenger's identity by airport staff. The system automatically generates labels to mark the luggage and the passenger hands it over to the check-in belt, Fig. 10.3. The luggage then passes through security. This step completes the check-in process. The advantage is faster passenger check-in without the need to wait in line and additional passenger identification and verification [15]. This system also relieves the burden on the airport attendants as the baggage check-in system is fully automatic and does not require the constant presence and attendance of airport staff.

After checking in the luggage, the passenger proceeds to passport control. Upon approaching the electronic gate, which is equipped with a camera sensor, the system records a photograph of the passenger's face. The passenger then places the passport

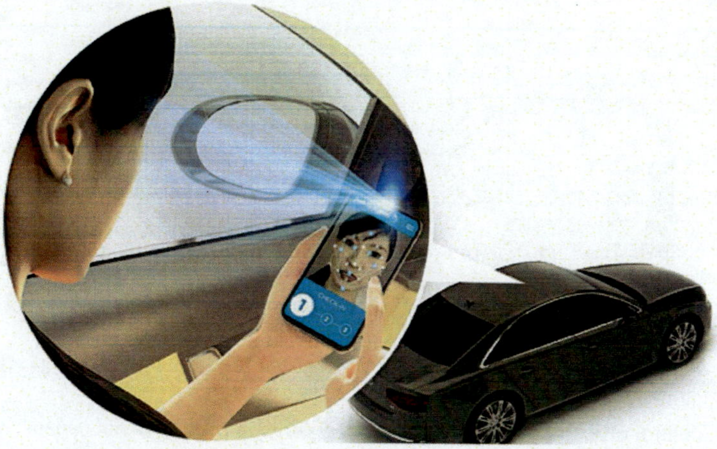

Fig. 10.3 Online check-in using biometrics [14]

Fig. 10.4 Baggage check-in using biometrics [14]

on the scanner, which captures the passport photograph. The system compares it with the facial features captured by the camera sensor; see Fig. 10.4.

In the case of a positive pairing, passengers can continue their journey. In case the system does not evaluate the match or the passenger is on the Dangerous Persons List, the passenger must undergo a personal passport check by the relevant passport

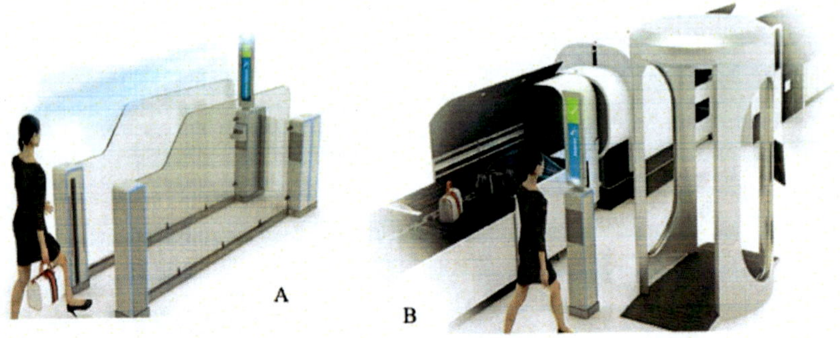

Fig. 10.5 **(a)** Passport control with facial recognition. **(b)** Security check using [14]

control officer. Since the biometric passport control gate is not dependent on human resources, the number of gates can be increased according to the airport's need without the need for staff augmentation [15]. With the use of biometric facial recognition technology, the time required to authenticate passengers will be significantly reduced, which will lighten the workload at the airport.

Automatic identification of passengers using AI at security check will help to speed up the identification process and consequently reduce the overall time to pass security check; see Fig. 10.5.

5 Conclusion

Solutions using facial biometrics are beginning to be implemented around the world. Long queues and high concentrations of people at boarding and passport control at airports are not ideal, especially in times of pandemics. Biometrics technology is bringing about major changes in the way companies operate across multiple industries and is playing a particularly large role in fundamentally transforming air travel. This paper focuses on the implementation of biometric technology in the Slovak Republic at the selected M. R. Štefánik International Airport in Bratislava.

Facial biometric technology can be applied in several areas within airport and passenger air transport, such as baggage check-in, passport control, security control, boarding, and access to restricted areas of the airport. The efficiency rate of facial biometric identification is sufficiently high for its use in air transport, as demonstrated by the growing number of airports implementing this technology.

Not only Airlines and airports but also other stakeholders in the commercial aviation industry should take proactive measures to enable them to implement solutions with facial recognition biometrics. Ultimately, facial biometrics will make the travel experience more efficient and enjoyable for the majority of passengers.

Biometrics refers to the development of statistical and mathematical methods that are applicable to data analysis problems in the biological sciences. It is also used to describe measurable physiological and/or behavioral characteristics that are innate and unique to an individual and that can be used to verify their identity. Therefore, from a research perspective, it is interesting to take a closer look at this topic and examine how selected consumer groups (e.g., age generations) view the use of biometric technologies in different areas.

References

1. Khan, N., & Efthymiou, M. (2021). The use of biometric technology at airports: The case of customs and border protection (CBP). *International Journal of Information Management Data Insights, 1*(2), 2021.100049, ISSN 2667-0968. https://doi.org/10.1016/j.jjimei.2021.100049
2. Burt, C. (2019). *Emirates Airlines biometric boarding with VeriScan at Dulles reaches full operation* [online]. Available from https://www.biometricupdate.com/201906/emirates-airlines-biometric-boarding-with-veriscan-at-dulles-reaches-full-operation. Accessed 15 May 2021.
3. Grünenberg, K., Møhl, P., Olwig, K., & Simonsen, A. (2020). Issue introduction: IDentities and identity: Biometric technologies, borders and migration. *Ethnos, 87*(2), 211–222.
4. Mir, U., Sharma, S., Kar, A., & Gupta, M. (2020). Critical success factors for integrating artificial intelligence and robotics. *Digital Policy, Regulation and Governance, 22*(4). https://doi.org/10.1108/DPRG-03-2020-0032
5. *SITA passengers in focus* [Electronic resource]. (2019). Available at:https://www.sita.aero/globalassets/docs/infographics/passengers-in-focus-infographic.pdf
6. *NEC's video face recognition technology ranks first in NISTTesting* [Electronic resource]. (2019). Available at: https://www.nec.com/en/press/201703/global_20170316_01.html
7. Negri, N. A. R., Borille, G. M. R., & Falcão, V. A. (2019). Acceptance of biometric technology in airport check-in. *Journal of Air Transport Management, 81*, 101720.
8. Kalakou, S., Psaraki-Kalouptsidi, V., & Moura, F. (2015). Future airport terminals: New technologies promise capacity gains. *Journal of Air Transport Management, 42*(C).
9. Haas, E. P. (2004). Back to the future – The use of biometrics, its impact of airport security, and how this technology should be governed. *Journal of Air Law and Commerce, 35*.
10. Filipi, C., dos Santos, G. (2022). *Gait recognition based on deep learning: A survey* [online]. https://arxiv.org/abs/2201.03323
11. WSJ. (2019). Are you ready for facial recognition at the airport? [online]. *The Wall Street Journal*. https://www.wsj.com/articles/are-you-ready-for-facial-recognition-at-the-airport-11565775008
12. *Types of biometrics*. Biometric Institute [online]. (2021). https://www.biometricsinstitute.org/what-is-biometrics/types-of-biometrics/
13. *Ways airports can use face recognition for safety and convenience* [online]. (2021). https://www.facefirst.com/blog/ways-airports-can-use-face-recognition-for-safety-and-convenience/. (2021).
14. Taguire, M. *The birth of biometric security* [online]. https://www.researchgate.net/publication/249376012_The_birth_of_biometric_security (2020).
15. *Comfort, safety, simplicity: How biometric identification is changing operations of airports* [online]. (2021). https://recfaces.com/articles/how-biometrics-changes-the-work-of-airports

Chapter 11
A New Trend in Car Personalization Based on Augmented Reality: A Study

Jozef Husár, Stella Hrehova, Lucia Knapčíková, and Justyna Trojanowska

1 Introduction

Augmented reality (AR) can create an immersive experience where we can see, hear and feel digital content that is just hidden from our physical world. With the help of AR, we can push the boundaries of reality to the very peak of experiencing experiences that can absorb us with their interactivity. Since it is an interactive tool that is financially costly and variable, the main pioneer in the development is the automotive industry. Considerable financial resources are invested in this area, and each car manufacturer comes with a number of different applications [1]. We should know that the main disadvantage of augmented reality is that it is based on the redistribution, collection and analysis of various types of data. They are taken as Big Data, which raises concerns about security and privacy rights [2]. Indeed, some augmented reality devices record the environment in real time. Recording can then raise potential legal concerns. Augmented reality shows us the thin reality between the real and the digital world. Another advantage, or rather disadvantage, is aligned with the possible dangers that result from the reality modification [3, 4]. A good example was the introduction of Pokemon GO, which caused a lot of controversy due to related accidents and even deaths. Overlaying digital elements on the natural environment hides real danger and thus reduces the concentration and caution of

J. Husár (✉) · S. Hrehova · L. Knapčíková
Technical University of Košice, Faculty of Manufacturing Technologies with a seat in Prešov, Department of Industrial Engineering and Informatics, Prešov, Slovak Republic
e-mail: jozef.husar@tuke.sk; stella.hrehova@tuke.sk; lucia.knapcikova@tuke.sk

J. Trojanowska
Poznan University of Technology, Faculty of Mechanical Engineering, Department of Production Engineering, Poznan, Poland
e-mail: justyna.trojanowska@put.poznan.pl

L. Knapcikova, S. Zohrehvandi (eds.), *9th International Conference on Mobility, IoT and Smart Cities*, EAI/Springer Innovations in Communication and Computing,
https://doi.org/10.1007/978-3-031-28225-6_11

users. Therefore, it is important to create certain standards for such cases. Developers should not overload their AR systems with digital elements. In addition, there is also a need to educate AR users so that they do not become addicted to AR to the point of becoming passive to real-world stimuli [5].

1.1 Areas of Use AR

AR technology is used in almost all fields, including art, business, archaeology, tourism, restoration, education, entertainment, leisure and industrial production, including the automotive industry [6]. In the field of tourism, augmented reality applications can help display various tourist attractions, restore historical places, or monuments with the help of mobile phone cameras, screen software and other technological means for a better idea to integrate real scenes. In addition to viewing different scenes, we can also obtain additional information [7]. Nowadays, AR technology is also often used in archaeology, where it is used to zoom in on relics in the real world to help archaeologists determine the origin, or more precisely the location, of their finds. AR technology allows consumers to see the product from every angle, without being forced to unpack the product. By scanning the image of the product with the help of applications for AR reality, you can better display some information or better see the appearance of the product [8, 9].

Repair and Maintenance

One of the biggest industrial use cases for AR is the repair and maintenance of complex equipment. Whether we are talking about a car engine or an MRI machine. Repair and maintenance workers use AR glasses and headsets to perform their jobs so they can work faster and more efficiently [10]. Through AR, they can be provided with useful information on the spot, suggestions for potential repairs, or point out potential problem areas. These examples of using AR in repairs are becoming more widespread as machine-to-machine IoT technology grows and can feed information directly into AR headsets [11, 12].

Business Logistics

AR presents many chances to increase efficiency and save costs in various areas of business logistics. Here, we are talking about storage, transport and route optimization. One of the world-famous shipping companies, DHL, has already incorporated smart AR glasses into some of its warehouses, where the lenses show workers the shortest path within the warehouse to find a certain item that needs to be shipped as quickly as possible. It offers workers more efficient ways to do their jobs and is among the best uses of return on investment in the business industry today [13].

Tourism

Technology has come a long way in recent years, with various changes and improvements in the travel industry, from information websites like Lonely Planet to review sites like TripAdvisor [14]. AR can thus provide many options for travel brands and agents, who will provide potential tourists with an even more realistic experience before the trip. Try to imagine that even before you book a ticket to New York, you will take part in a virtual "Walkabout" of the USA on augmented reality glasses, or go for a walk through the streets of Rome, so that you can imagine what sights you can visit, or in which you can eat in restaurants. AR promises that the sale of trips, travel and holidays will be much easier and more accessible in the future [15].

Design and Modelling

From interior design to architecture and construction, AR helps professionals visualize their final products during the creative process. Thanks to the use of headsets that allow engineers, architects and design professionals to enter spaces and buildings before they are built [16]. They can thus better imagine how their plans look and can even make virtual changes to their designs on the spot. City planners can even model what the layout of an entire city might look like using an AR headset visualization. All modelling and design work that involves spatial relationships is a perfect use case for AR technology [17].

Retail

In today's real world, customers are increasingly using their smartphones to make purchases. Thanks to them, customers can find out more information about the given product or compare the price with other sellers. A very good example is the world-famous brand for the sale of Harley Davidson motorcycles. This brand decided to make the most of this trend and thus created its own AR application, which customers can use directly in the store on their smartphones. Thanks to it, customers can view the motorcycle they are interested in directly in the store to get a better idea or simply configure their own motorcycle according to their own ideas with their own colours or accessories, which will be displayed to them with the help of the application directly in the store, before it is manufactured [18].

1.2 Sale

When selling a vehicle, each salesman has to deal with different customer requirements. He will want to present as many models as possible with as many equipment options as possible to the potential customer, but we often find only one or two

Fig. 11.1 Vehicles sale in a car showroom

models in the showroom. Even if the desired model is accessible, many things cannot be simply observed, such as braking distance or turning radius. One way to solve this problem is to use VR devices [19]. However, working with these devices has its own drawbacks. The view using VR is limited because the volume and dimensions of the vehicle are distorted in the user's perception. To solve these problems, we use augmented reality, which offers us interesting alternatives. As with VR, we can also customize the car interactively with AR. When using AR devices, we achieve an increase in the real vehicle, while its volume and dimensions are preserved [20]. With AR, we can improve the customer experience and show them a better idea with the help of interactions that take place between the real environment and virtual elements. For example, if the customer would like a different colour of the bodywork, restitution will occur during which the current light conditions will be preserved, but on the other hand, the reflections directed from the surrounding environment to the bodywork will remain unchanged. AR technology can better visualize the equipment elements for the average customer and more so for commercial vehicles when their interior equipment is laid out [21] (Fig. 11.1).

Augmented reality is used as a sales assistant. Until now, the limited VSALM could accurately align the CAD model of the car with the video (top left); augmented reality can also be used to select the interior (bottom left) of the car. The entire processing takes place in real time on a mobile tablet (bottom right).

2 Research

In this section, we prioritize augmented reality as a marketing tool. The progress of automobile companies is related to the use of the most modern technologies. In recent years, experts from the automotive industry have increasingly come across the use of technologies in the automotive industry in their work. Augmented reality helps in various aspects of the automotive industry, such as car assembly and maintenance, prototyping, showrooms and sales itself [22].

Two types of augmented reality applications are most widely used in the automotive industry. The first type is sensor-based applications and the second marker-based applications. Applications that are based on markers are used in car showrooms. With this application, marketing teams can present the higher potential of a new product. Marker-based AR applications are mainly used for training purposes [23, 24]. Engineers and designers of the automotive industry use it during implementation to proceed correctly according to the correct instructions and schedule. Sensory applications will provide the user with a sense of false presence in an alternative reality. This technology is mainly based on sound, visual and multimedia stimuli, so that the user gets the feeling that he is somewhere completely different [25]. In the automotive industry, sensory applications using augmented reality are used to test vehicles before their production or to test the safety features of a new vehicle model by simulating a dangerous situation that could occur. A great example of sense-based augmented reality is the LiDAR security system. It calculates the distance between the car and objects and estimates the projected collision time accordingly [26].

2.1 *Reduction of Design and Prototyping Time and Costs*

The biggest problem in designing and creating prototypes has always been creating physical models. In this case, augmented reality works by overlaying physical objects with new elements. Car manufacturers use spatial augmented reality to project certain images onto physical car models. Using a projector is a cheap and convenient way to integrate augmented reality into the process of prototyping and designing car models, compared to a physical model factory. Automotive companies use glasses with augmented reality during development, which brings not only higher costs than using a projector but also more accurate and wider possibilities of use. With the help of augmented reality, the work of designers is much easier, and changes can be made directly during the discussion and are visible immediately, for all participants. With the use of augmented reality technologies, the duration and cost of developing a new design are reduced [10].

2.2 *Optimizing the Assembly Process*

Assembling a car requires absolute precision and attention to detail. During the assembly process, experienced operators also check the technical documentation. Checking such technical documentation significantly slows down the assembly process and can cause multiple errors. That is why augmented reality was used for the assembly process. All information regarding the work line is provided for operators in the form of images, videos and instructions, where they are explained step by step on the screen of smart glasses with augmented reality. Various functions can be created by employees using voice control. With the use of augmented reality smart glasses, the assembly process of the production line was significantly accelerated and assembly errors were minimized [15].

2.3 *Digital Car Showrooms*

If you decide to buy a new car today, you no longer need to visit a brick-and-mortar car showroom to view the car. All you have to do is download the augmented reality application of the given brand to your smartphone. In 2018, the world-famous car brand Porsche released the Mission E Augmented Reality application, which you can download for free on your smartphone. The company created this application mainly for the sale of new Porsche electronic vehicles. In EXPLORE mode, you can create your own 3D design. In the next DRIVE mode, you can try a test drive in augmented reality. Such types of car demos primarily save costs and can provide interested parties with everything they are looking for in a given model [20].

3 Research Methodology

Some car brands offer us the configuration of our own car directly from the comfort of our home. You can create a likeness of the car you plan to buy using the app of the given brand. Not all car brands have freely available applications, but there are a few. If you do not find your brand freely available on the Internet, this option will be certainly provided you in the car manufacturer showroom. The use of these applications is often tied to the operating system, where more car manufacturers offer options for IOS than for ANDROID. The greater reach of these technologies is more freely available for the Western market, where these applications have great success.

3.1 *AR VISUALISER Porsche*

The Porsche AR VISUALISER application is available for iOS operating systems, where you can choose from more than 20 models from this premium brand (Fig. 11.2).

The application allows us to visualize the car in real size. We can view the car from every side and from different angles. We can look at any part of the car in more detail. Thanks to the spatial application, we can also enter the interior of the car. Inside, we see the design and storage of all vehicle elements as in a real car. It is

Fig. 11.2 Porsche Carrera 911 – AR VISUALISER (own processing)

possible to monitor the space of the seats both in front and in the back. The application panel offers us to change the colour design of the vehicle "colours." There are six types of colour shades to choose from in which the given model is actually produced. In the toolbar, there is an option to replace "wheels" discs. There are more than 10 types of different discs and sizes in the options. The colour of the discs is not arbitrary, and we can only choose from the discs that Porsche offers, both the discs and their colour combination, which is given in advance. For each disc, the manufacturer indicates their size, colour and shade, whether it is a glossy or matte type of surface. The application also offers us the opportunity to test of car drive. After starting "drive mode," the icons D (drive) will appear on the screen, driving in the front, R (reverse) driving, Brake as a brake. Thanks to movements of the mobile phone to the right or left side, the car can turn. This is how you can easily test your vehicle during an online test drive. In my opinion, this is a very useful step if you want to imagine how your chosen discs will look in motion of the car (Fig. 11.3).

3.2 *ToyotaARShowroom*

The ToyotaARShowroom application offers us five car models for our Slovak market, where you can configure the given model for free. The application version is available for IOS operating systems (Fig. 11.4).

ToyotaARShowroom offers a presentation in three scales 1:1, 1:50 and 1:25. The car can be viewed in 360° from every side, so you can view every detail perfectly. On the main page, we see three available icons with options for selecting given changes. Toyota offers us a change of colour only in three selected shades. They are combinations of white platinum with a black roof, red with a black roof and coral red with a black roof. Each colour combination is also marked with an exact code. The

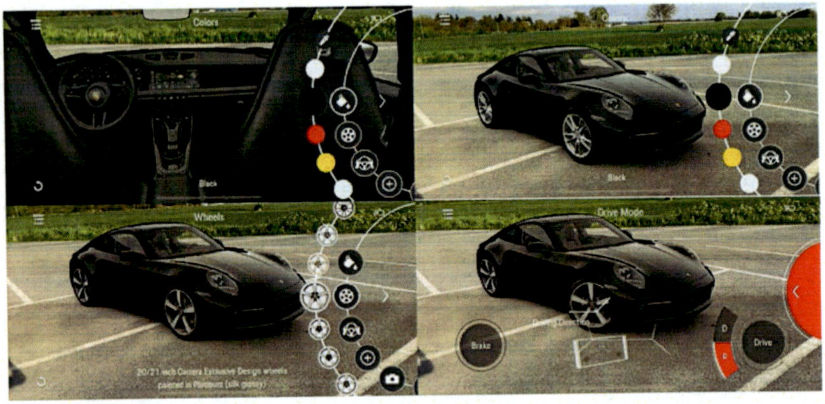

Fig. 11.3 Variants of the application Porsche AR VISUALISER (own processing)

Fig. 11.4 Toyota Yaris – ToyotaARShowroom (own processing)

choice of colour variants is limited for the application; more colour change options can only be done directly at the seller. In the application, there is also the option of choosing the motorization for the given model. The Toyota Yaris model is produced with a 1.5 Dynamic Force gasoline engine with a capacity of 125 hp, where there is also a choice between a manual six-speed transmission and an automatic transmission. There is also a hybrid version of this model with an output of 116 hp in conjunction with an automatic transmission. The drive of all motors is always only 2×4 (Fig. 11.5).

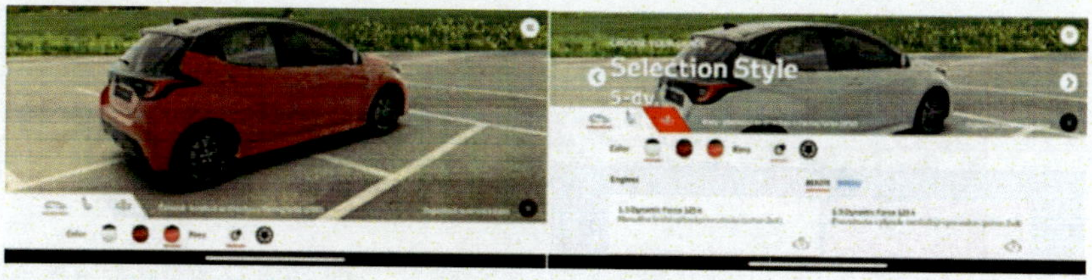

Fig. 11.5 Variation of application ToyotaARShowroom (own processing)

Fig. 11.6 Lexus IS 350 – Lexus AR Play (own processing)

3.3 *Lexus AR Play*

The Lexus AR Play application offers only one model for the Slovak market, the Lexus IS350. The application is only available for devices with the IOS system (Fig. 11.6).

We can view the model in any size. The picture shows the model in real size, which can be viewed from all sides, both exterior and interior. On models, it is possible to open the driver's or passenger's door. A look inside offers us a view of the red interior, the colour version of which is not possible in the application. The display of interior elements corresponds to the real layout of things in relation to the real model. It is possible to choose a colour design from eight variants. However, the names of the colours are not described and their performance may be distorted. In the design option, blue parts are displayed on the models, which indicate the possibility of editing the given model. This is a replacement of parts for sports accessories. However, the accessories are not displayed in the application, and for their implementation, it is necessary to visit an authorized store, where workers can use their devices to create an accurate image. On this model, there is the option of choosing a sports hood, sports wheel arches, a sports rear bumper diffuser and also changing the

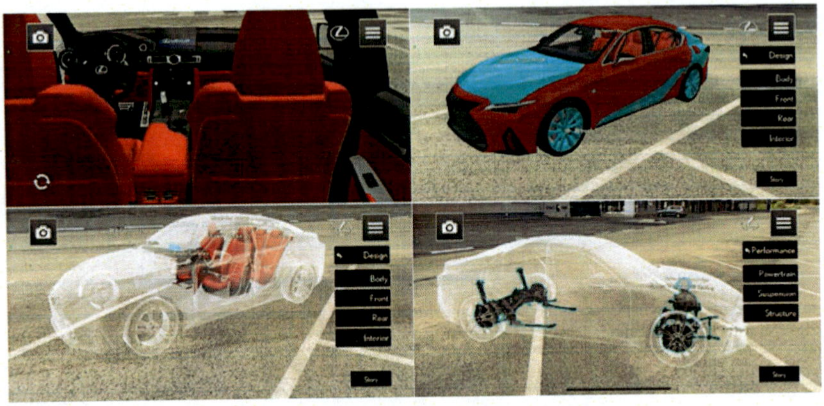

Fig. 11.7 Variants of the application Lexus AR Play (own processing)

size of the discs. In the BODY section, we will see the headlights together with the front grille of the bumper. There is a choice of headlights, either halogen or LED. We have the option to change the grille of the front mask to the sporty look that Lexus offers. However, these operations are not available for the app, and you need to visit an authorized Lexus dealer to view them. Thanks to the BODY option, where only the silhouette of the car's skeleton remains and only the interior is fully displayed, we can view the placement of the seats and dashboard. On the instrument panel, we will see the data of the car radio, where Lexus offers both Android and Apple connectivity for content sharing. The car radio has a 10.3-inch touch screen. In the PERFORMANCE section, we are left with only the silhouette of the car's frame, where parts of the exhaust pipe will be shown along with the complete set-up of the engine and gearbox. This possible visualization is especially suitable for mechanics and people who are interested in the operation of the engine. The truly car heart, namely the engine, can thus be viewed in complete detail without having to dismantle anything. With the help of the so-called X-ray, we can see the individual parts that are hidden under the bodywork. We can see the placement of the engine, gearbox or axles (Fig. 11.7).

4 Results

In the article, we focus on three well-known car brands of different price categories. We first focused on the brand of Porsche sports cars, which belong to a higher price category. Porsche cars stand out for their design combined with quality and a powerful sports engine. We meet this brand of cars quite rarely on the roads. The occurrence of these models is primarily in more developed countries. In terms of price, these models range in higher prices, which means that the configuration using augmented reality helps to better imagine and satisfy every customer, because the materials that are used in the construction reach astronomical prices. In such a case, it is very financially demanding to assemble several models types in order to be able to satisfy every customer.

Toyota cars are increasingly rolling the automotive industry with their models, where their reliability and timelessness are often discussed. Toyota combines conventional gasoline or diesel engines with electric ones. The choice of these options is indeed large, and the company wants to show its customers the maximum.

Lexus is one of the lesser-known brands in our region, but in the Western Hemisphere, they are very popular cars that combine luxury with sporty driving. The quality of these cars is at a very high level, which is why this brand is chosen by the biggest car enthusiasts (Table 11.1).

In the table, we have evaluated the most important parts offered by the applications. Each application offers several options. However, the applications we compared are limited to the Slovak market, so there is no possibility of completely detailed changes. For a completely perfect configuration, it is already necessary to visit an authorized service, where the sellers will configure the car for you completely according to your wishes.

Table 11.1 Comparison of AR applications

Functions	Porsche AR	Toyota AR	Lexus AR
Available for the Slovak market	Yes	Yes	Yes
Operating system IOS	Yes	Yes	Yes
Operating system Android	No	No	No
View from all sides – 360°	Yes	Yes	Yes
Changing the vehicle colour	Yes	Yes	Yes
Detailed information about the colour	No	Yes	No
Electron selection	Yes	No	Yes
Choice of engine	No	Yes	No
Gear selection	No	Yes	No
Test drive	Yes	No	Yes
Engine sound effects	Yes	No	Yes
Turning on the lights	Yes	No	No
Opening the door	Yes	No	Yes
Sports modifications	No	No	Yes
X-ray of the car	No	No	Yes
A wealth of options/functions	2.	3.	1.
Application language	EN	SK	EN
Graphic design	1.	2.	3.
Clarity of the application	1.	2.	3.
Number of vehicle models	20	5	1

5 Conclusion

In the article, we pointed out three completely different types of applications. From the analysed point of view, the application from Porsche took the first place. The application is easy to use, there is a sufficiently large selection of vehicles, and basic things related to configuration are available for free. Advantage is a rich selection of electrons and a well-crafted graphic design. The Toyota AR app offers few configuration options. You can play the car in the space in real size, but fundamental changes are not available for free and you need to visit the store. The Lexus AR application offers only one model for the Slovak market. The availability of configuration options is wide, but specific modifications are only indicated and cannot be performed; also it has a weaker graphical visualization. The applications are suitable for a better idea of the vehicle in a real environment, but for a complete configuration, it is always necessary to visit an authorized dealer who has unlimited capabilities in the application.

Acknowledgements This work was supported by the Slovak Research and Development Agency under the contract No. APVV-19-0590, by the projects VEGA 1/0268/22, KEGA 038TUKE-4/2022 granted by the Ministry of Education, Science, Research and Sport of the Slovak Republic.

Paper is the result of the Project implementation: Development of excellent research capacities in the field of additive technologies for the Industry of the twenty-first century, ITMS: 313011BWN5, supported by the Operational Program Integrated Infrastructure funded by the ERDF.

References

1. Buń, P., Trojanowska, J., & Rewers, P. (2019). VR and AR in lean manufacturing classes. In J. Trojanowska, O. Ciszak, J. Machado, & I. Pavlenko (Eds.), *Advances in manufacturing II. Manufacturing 2019* (Lecture notes in mechanical engineering). Springer. https://doi.org/10.1007/978-3-030-18715-6_29
2. Dunleavy, M. (2014). Design principles for augmented reality learning. *TechTrends, 58*, 28–34. https://doi.org/10.1007/s11528-013-0717-2
3. Kascak, J., et al. (2019). Implementation of augmented reality into the training and educational process in order to support spatial perception in technical documentation. In *IEEE 6th international conference on industrial engineering and applications (ICIEA)* (pp. 583–587). IEEE. https://doi.org/10.1109/IEA.2019.8715120
4. Trojanowski, P., & Trojanowska, J. (2021). Reliability of road transport means as a factor affecting the risk of failure – The transport problem case study. In V. Ivanov, J. Trojanowska, I. Pavlenko, J. Zajac, & Perakovi'c, D. (Eds.), *DSMIE 2021. LNME* (pp. 253–261). Springer. https://doi.org/10.1007/978-3-030-77719-7_26
5. Husár, J., Kaščak, J., & Hrehová, S. (2022). Use of marker AR by implementation on production equipment for education of students of technical fields: A study. In L. Knapcikova, D. Peraković, M. Perisa, & M. Balog (Eds.), *Sustainable management of manufacturing systems in industry 4.0* (EAI/Springer innovations in communication and computing). Springer. https://doi.org/10.1007/978-3-030-90462-3_7
6. Husár, J., & Knapcikova, L. (2021). Possibilities of using augmented reality in warehouse management: A study. *Acta Logistica, 8*(2), 133–139. https://doi.org/10.22306/al.v8i2.212

7. Židek, K., Maxim, V., Pitel, J., & Hosovsky, A. (2016). Embedded vision equipment of industrial robot for inline detection of product errors by clustering-classification algorithms. *International Journal of Advanced Robotic Systems, 13*, 1–10. https://doi.org/10.1177/1729881416664901
8. Himperich, F. (2007). *Applications in augmented reality in the automotive industry*. Fachgebiet Augmented Reality, Department of Informatics, 1–21.
9. Ivanov, V., Botko, F., Dehtiarov, I., Kočiško, M., Evtuhov, A., Pavlenko, I., & Trojanowska, J. (2022). Development of flexible fixtures with incomplete locating: Connecting rods machining case study. *Machines, 10*, 493. https://doi.org/10.3390/machines10070493
10. Zhylenko, T., Ivanov, V., Pavlenko, I., Martynova, N., Zuban, Y., & Samokhvalov, D. (2022). Mobile applications in engineering based on the Technology of Augmented Reality. In J. Machado, F. Soares, J. Trojanowska, & S. Yildirim (Eds.), *Innovations in mechatronics engineering. ICIENG 2021*. Lecture Notes in Mechanical Engineering. Springer. https://doi.org/10.1007/978-3-030-79168-1_33
11. Kujawińska, A., & Diering, M. (2021). The impact of the organization of the visual inspection process on its effectiveness. *The International Journal of Advanced Manufacturing Technology, JAMT-D-20-02941R1, 112*, 1295–1306. https://doi.org/10.1007/s00170-020-06543-9
12. Varela, M. L. R., Putnik, G. D., Manupati, V. K., et al. (2021). Integrated process planning and scheduling in networked manufacturing systems for I4.0: A review and framework proposal. *Wireless Network, 27*, 1587–1599. https://doi.org/10.1007/s11276-019-02082-8
13. Gay-Bellile, V., Bourgeois, S., & Tamaazousti, M. (2015). Applications of augmented reality for the automotive industry. In *Fundamentals of wearable computers and augmented reality*.
14. Antosz, K., & Stadnicka, D. (2018). An intelligent system supporting a forklifts maintenance process. In A. Burduk & D. Mazurkiewicz (Eds.), *Intelligent Systems in Production Engineering and Maintenance – ISPEM 2017. ISPEM 2017. Advances in intelligent systems and computing* (Vol. 637). Springer. https://doi.org/10.1007/978-3-319-64465-3_2
15. Trojanowski, P. (2020). Comparative analysis of the impact of road infrastructure development on road safety – A case study. *Scientific Journals of the Maritime University of Szczecin, 63*(135), 23–28. https://doi.org/10.17402/436
16. Boboc, R. G., Gîrbacia, F., & Butilă, E. V. (2020). The application of augmented reality in the automotive industry: A systematic literature review. *Applied Sciences, 10*, 4259. https://doi.org/10.3390/app10124259
17. Grześkowiak, M., & Trojanowska, J. (2021). Production line balancing in a mixed-model production system: A case study. In V. Tonkonogyi et al. (Eds.), *Advanced Manufacturing Processes II. InterPartner 2020* (Lecture notes in mechanical engineering) (pp. 24–32). Springer. https://doi.org/10.1007/978-3-030-68014-5_3
18. Demčák, J., Lishchenko, N., Pavlenko, I., Pitel', J., & Židek, K. (2022). The experimental SMART manufacturing system in SmartTechLab. In J. Trojanowska, A. Kujawińska, J. Machado, & I. Pavlenko (Eds.), *Advances in manufacturing III. Manufacturing 2022* (Lecture notes in mechanical engineering). Springer. https://doi.org/10.1007/978-3-030-99310-8_18
19. Häuslschmid, R. (2018). *Extending head-up displays: Exploring the potential of large & 3D automotive windshield displays* (Dissertation). LMU München. https://doi.org/10.5282/edoc.22766
20. Kaščak, J., Husár, J., Knapčíková, L., Trojanowska, J., & Ivanov, V. (2022). Conceptual use of augmented reality in the maintenance of manufacturing facilities. In J. Trojanowska, A. Kujawińska, J. Machado, & I. Pavlenko (Eds.), *Advances in manufacturing III. Manufacturing 2022* (Lecture notes in mechanical engineering). Springer. https://doi.org/10.1007/978-3-030-99310-8_19
21. Kawęcki, N., & Gola, A. (2022). Pick performance system as an IT support for order completing – A case study. In J. Trojanowska, A. Kujawińska, J. Machado, & I. Pavlenko (Eds.), *Advances in manufacturing III. Manufacturing 2022* (Lecture notes in mechanical engineering). Springer. https://doi.org/10.1007/978-3-030-99310-8_9

22. Trojanowski, P., Trusz, A., & Stupin, B. (2022). Correlation between accidents on selected roads as fundamental for determining the safety level of road infrastructure. In V. Ivanov, J. Trojanowska, I. Pavlenko, E. Rauch, & D. Peraković (Eds.), *Advances in design, simulation and manufacturing V. DSMIE 2022* (Lecture notes in mechanical engineering). Springer. https://doi.org/10.1007/978-3-031-06025-0_11
23. Kluz, R., & Antosz, K. (2019). Simulation of flexible manufacturing systems as an element of education towards industry 4.0. In J. Trojanowska, O. Ciszak, J. Machado, & I. Pavlenko (Eds.), *Advances in manufacturing II. Manufacturing 2019*. Lecture Notes in Mechanical Engineering. Springer. https://doi.org/10.1007/978-3-030-18715-6_28
24. Kłosowski, G., Gola, A., & Thibbotuwawa, A. (2018). Computational intelligence in control of AGV multimodal systems. *IFAC-PapersOnline, 51*(11), 1421–1427. https://doi.org/10.1016/j.ifacol.2018.08.315
25. Antosz, K., Pasko, L., & Gola, A. (2019). The use of intelligent systems to support the decision-making process in lean maintenance management. *IFAC PapersOnLine, 52*(10), 148–153. https://doi.org/10.1016/j.ifacol.2019.10.005
26. Hrehová, S., Husár, J., & Knapčíková, L. (2022). The fuzzy logic predictive model for remote increasing energy efficiency. *Mobile Networks and Applications*. https://doi.org/10.1007/s11036-022-02050-1

Chapter 12
Management and Maintenance of Buildings in BIM Environment: Facility Management of Historic Buildings (HBIM)

Peter Mésároš, Mária Grazianová, Jana Smetanková, Katarína Krajníková, and Annamária Behúnová

1 Introduction

When we talk about managing a successful project, we must consider that the magic formula lies in having a complete and detailed notion of the resources or information with which you work. In the construction sector, especially with existing buildings, this step may seem more critical, because with already built buildings, the level of know-how and accessibility to data may be lower [1, 2].

In order to obtain an effective, efficient and safe operating strategy in the maintenance management (MM) phase, everything that comes from the design and construction phases is essential [3].

In the hope of reckoning the costs of proper interoperability in the facility industry, the National Institute of Standards and Technology (NIST) recreates through an analysis conducted in the USA, a timeless model and a reliable source

P. Mésároš · M. Grazianová · K. Krajníková
Technical University of Kosice, Faculty of Civil Engineering, Institute of Construction Technology, Economics and Management, Kosice, Slovakia
e-mail: peter.mesaros@tuke.sk; maria.grazianova@tuke.sk

J. Smetanková (✉)
Technical University of Kosice, Faculty of Civil Engineering, Expert's Institute in Construction, Kosice, Slovakia

National Infrastructure for Technology Transfer Support in Slovakia II – NITT SK II, Bratislava, Slovakia
e-mail: jana.smetankova@tuke.sk

A. Behúnová
Technical University of Košice, Faculty of Mining, Ecology, Process Control and Geotechnologies, Institute of Earth Resources, Kosice, Slovakia
e-mail: annamaria.behunova@tuke.sk

L. Knapcikova, S. Zohrehvandi (eds.), *9th International Conference on Mobility, IoT and Smart Cities*, EAI/Springer Innovations in Communication and Computing,
https://doi.org/10.1007/978-3-031-28225-6_12

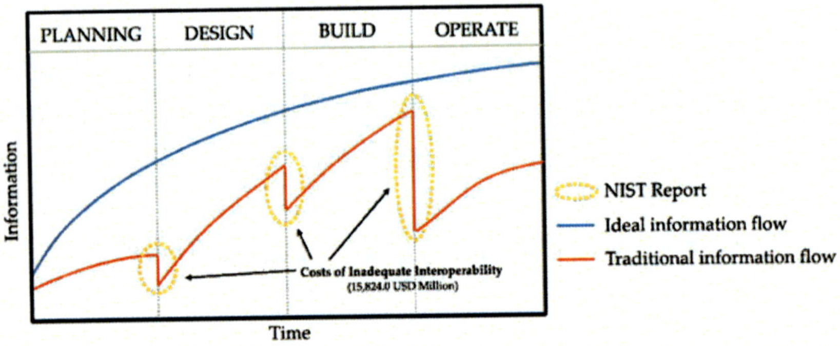

Fig. 12.1 Loss of information in the building life cycle and cost estimation in the NIST study. (Rework based on Eastman et al. [5] and Gallaher et al. [4])

of data: inadequate interoperability costs of capital facility projects in the USA in 2002 amounted to a disproportionate total of $15.8 billion, out of the approximately $10.6 billion accounted for by facility management (FM) owner/participants [11].

Going to do an analysis of data processing between phases of a building's life cycle (red line), we traditionally find an efficient data loss between them (Fig. 12.1). Comparing everything with the actual workflow (blue line), we note that, to minimize the gap between the two (red-blue) and at the same time, reduce the information gap that gradually becomes evident, it is very effective to establish a methodological framework adequate, also thanks to digitization or more powerful systems [4].

Based on the facts, new innovative ways and methods of management and maintenance of buildings, respectively, facility management of buildings are, therefore, increasingly being developed. The detail, availability and completeness of data provide an important tool to support the effective management of facility management of buildings. In the case of buildings built in the past (historical buildings), this information is difficult to access, incomplete and of insufficient quality. Building information modelling (BIM), or heritage BIM (HBIM), is a new way of preparing, planning and managing facility management for historic buildings, the application of which in practice needs to be monitored, constantly supported and increase its rate of implementation.

2 Literature Review

2.1 *Facility Management (FM)*

Facility management is an organizational function that integrates people, processes and place into the built environment. The priority goal is to improve the quality of life of people and the productivity of the main activities [6].

Facility management begins after the construction is completed and the building is handed over. It is a continuous process throughout the entire construction life cycle. The main goal is to ensure the management of the property and its contents (equipment, equipment, premises, etc.). A building cannot function without the absence of facility management, as it includes all activities integrating business management, asset management, maintenance, contract management, renovation, etc. [7].

In 1990, Becker defined facility management as a discipline responsible for coordinating all efforts related to the planning, design and management of a building, its systems, equipment and furniture in order to increase the organization's capability [8].

Moore and Finch (2004) defined facility management as the development, coordination and management of all ancillary specialist services of an organization, together with buildings and their systems, plant, IT equipment, fixtures and fittings, with the overall aim of helping any organization achieve its strategic objectives [9].

The aim of the mentioned operations within facility management is to create an environment in which systems work together smoothly. The goal of the facility manager is that all components of the building work properly. The facility manager ensures that the systems in the built environment work together to fulfil their intended purpose and that the staff is healthy and productive. The main goal of the facility manager is to compile a list of tasks, primarily:

- Cleaning, safety, maintenance and land management plan
- Processes and principles of work
- Emergency (emergency/catastrophic) plan and mitigation of its impacts
- Sustainability planning
- Project management and budgeting
- Property management and space planning
- Business continuity planning [6]

Facility managers contribute to the creation of organizational strategies and their results are as follows:

- Proposals to improve operational efficiency
- Planning and delivery of infrastructure needs to support productivity
- Risk management, including risk management of facilities, employees, suppliers and trade name
- Mitigation and reduction of environmental impacts
- Supporting sustainable tactics and long-term cost management
- Use of technological solutions
- Mitigation and elimination of the effects of natural disasters
- Ensuring compliance
- Use of security [6]

The basic competencies of the facility manager are as follows:

- Leadership and strategy
- Operations and maintenance
- Finance and business
- Sustainability
- Communications
- Occupancy and human factors
- Performance and quality
- Facility information management and technology management
- Real estate
- Risk management [10]

2.2 Building Information Modelling (BIM)

Building information modelling (BIM) represents the process of digital representation of the functional and physical characteristics of a building, through a digital model and its individual dimensions. BIM represents a holographic approach to designing the construction, operation and maintenance of a building project as a collaborative process, which uses a unified coherent system of digital modelling. BIM improves design visualization including digital simulations and tests across all phases of the building life cycle (planning, design; construction, implementation; and operation, facility management).

The main aim of the building information modelling approach is to guarantee the exchange of information during the entire life cycle of the building and to create a coherent and organized information model. The power of BIM lies in the ability to define a real data model of the building, which provides not only graphics but also additional information that was gradually introduced during the development – life cycle of the building. This type of information is the basis for setting the right framework for effective management.

The implementation of BIM has brought changes in the architecture, engineering and construction industry (AEC). BIM supports the introduction of new ways of designing, building and managing buildings. Building information modelling tools help collaboration between individual participants, enable digital management and access to information at any time [11].

BIM modelling improves efficiency in construction companies. When working in a BIM environment, it is possible to predict and point out many possibilities and potential problems, and at the same time generate a lot of valuable information that can reduce the amount of time needed for important operational projects.

In the design phase and construction planning phase, BIM is used to model the proposed building. Architects and designers use BIM for visualization and planning. With the help of BIM modelling, the user gets a database of real information, designers and project managers know how the space will be used (they create and run various simulations, etc.).

Valuation – cost estimation is a long and difficult process. Incorrect determination of the acreage statement is the main shortcoming when determining the construction budget. BIM works not only as a 3D representation of the building but also as a database of parts. When determining the statement of dimensions, it is possible to use models created in the BIM environment, which allows us to obtain a very accurate estimate of individual constructions, materials and products.

Project managers can use BIM modelling to increase the level of coordination and eliminate harmless clashes and collisions. By checking the BIM model or running a simulation, designers can prevent clashes and collisions early.

Another advantage of using building information modelling is the improvement of documentation quality. With an overview of coordination, possible conflicts and an overview of planning, the project manager will obtain important information to improve project documentation (more accurate estimates, more accurate numbers, elimination of errors and collisions) and improve cooperation between individual participants. Improving documentation and reducing the time required to prepare documents improve planning. Project managers have access to the model, the timeline, and with the help of current variables, they know the exact construction duration. With these data, it is possible to achieve better coordination of the work schedule and thus improve on-site planning. Project managers can use BIM modelling and their experience to provide more accurate project schedules. A better schedule will improve operational efficiency within construction [11, 12].

2.3 *Building Information Modelling and Facility Management*

BIM plays an important role in providing accurate information about the design of the building and other elements

BIM represents a holistic approach to the design, exhibition and management of buildings and facilities. Currently, BIM tools are used primarily in the design and construction of buildings. The model created in the BIM environment (in the design and construction phase) represents an important tool in providing accurate information about the design of the building and other elements to the building manager or owner who can then effectively manage the life cycle of the building with this data. BIM makes it possible to create a central database that can be referenced and used as a basis for maintaining the life of the building. BIM therefore provides a complete database of individual structural parts, floor plans, spatial requirements, equipment, electrical installations and the like. BIM is primarily used in the field of facility management in the following areas:

- Property management: with correct data on property, floor plans, space requirements, it is possible to eliminate wastage of space, which leads to an extraordinary reduction in real estate expenses. Information obtained through BIM tools is the basic data for the efficient use of the building's premises.

- Preventive maintenance planning: the main goal of planned maintenance is to streamline maintenance processes and create programs by entering information about assets, building structures, electrical components and the like. This information provides administrators with data to plan maintenance, perform repairs and scope depending on warranty, service requirements, and maintenance manuals. This useful information about construction equipment is stored in BIM models and can easily reduce the tedious processes of filling out maintenance records,
- Efficient energy use: energy use has a direct impact on the environment and is reflected in operating costs. It is very important to use the building's energy efficiently. BIM helps the architect of the building in the cost analysis and makes it possible to compare different alternatives to reduce energy costs. The administrator can continuously monitor the performance during the life of the building and other important factors such as comfort, water consumption, electricity consumption and the like. BIM provides the necessary information and helps in formulating the best energy consumption strategy with the minimum possible costs,
- Of additional equipment, reconstructions and renovations: the BIM model of the building provides an overview and basis for future planning of modernization, reconstruction and renovation. BIM helps facility managers and owners to make better decisions based on sufficient data about the existing building. This simplifies the process and minimizes costs. By providing detailed and accurate data, the costs of complex renovations of buildings and modernization of projects will be reduced, thereby eliminating a lot of inter-disciplinary clashes (technical equipment of buildings vs. structural solutions), which usually occurs,
- Improvements in building life cycle management: building life cycle data is very important and necessary. Thanks to the complexity of the data from the design, construction to its implementation, the owner gets information about the expected life and replacement costs. Thus, BIM models help the building manager to analyse the benefits of investing in materials and systems that can significantly reduce one-time costs or prove to have a long-term effect, thus providing a better return during the life cycle of the building [11].

2.4 *Heritage Building Information Modelling (HBIM)*

Unlike new build projects, in the cultural heritage sector we see that the BIM scope is quite a new academic field of exploration. The dissimilarity and therefore the difficulty of standardization are in this regard the challenge to be overcome with older buildings. Traditional asset management methods do not guarantee digital data on a single digital platform. Working in a multi-format system, documents such as PDFs, Word reports, Excel forms, etc., have difficulty relating and linking. Furthermore, being governed by a two-dimensional representation system, it is missing in the third dimension and, therefore, it is difficult for managers to make serious choices. In such a structure, obtaining and updating data are therefore difficult.

In addition, data exchange between stakeholders is difficult because different formats create a lack of collaboration and communication gap. Adopting such a traditional method causes many problems during manipulation, reuse and FM process. Data exchange between interested parties is difficult, because different formats create a lack of cooperation and a communication gap.

The presence of different formats determines a lack of participation and a divergence of dialogue such that the exchange of data between the members involved is difficult. Thus, adopting a traditional system creates several problems in the handling, reuse and FM processes. The introduction of the concept of BIM in the heritage field saw its beginnings in 2009 under the name of Heritage Building Information Modelling (HBIM) [13].

When we talk about HBIM, we are talking about a built heritage management system aimed at defining an integrated three-dimensional model and with its historical data. The importance of this system is manifested in the integration of historical documents, information about the construction, monitoring and status in use of the building in the 3D area [14]. Lately, the scope of application of BIM on historical buildings has been re-encountered in BIM scanning and parametric reconstruction [15, 16]. Anyway, this area is still little explored if we look at the systems of integration of semantic information particularly related to the management of assets and the life cycle in other times with complex forms and its application to the planning of reconstruction activities and FM. To make valuable decisions on the life cycle of aging buildings, it is imperative for owners, managers and organizations to find application of HBIM in the integration of semantic information and its use in historical management.

From what has been highlighted, it is possible to note the imminence of developing and demonstrating an HBIM scheme capable of retrieving documents and managing historical data from multiple sources in a single BIM environment in order to use it for archival purposes and in reuse planning, of restoration and FM.

Lately, HBIM has seen application in the modelling, documentation, conservation and sorting of historic architectural characteristic [13, 17]. Creating a historical model is possible using graphical data obtained from 3D laser scanning and photogrammetry [17]. Thanks to the laser scanning method, the surfaces of the historical structures are created in a 3D environment by means of a series of selected points which indicate their geometric coordinates [18]. The laser beam of the scanner focuses on a target and its reflection returns the precise structure of the heritage [19]. Photogrammetry, instead, gives information on the actual state of surfaces thanks to photographic images. The images taken by the camera through selected points on the area (GCP) are then developed in the structure for motion (SfM) algorithm [20]. The result of all this is a detailed geometric representation of the 3D elements obtained from point clouds composed of an enormous number of points with geometric coordinates [21]. The next step aimed at preserving the articulated structure of the artefact is preceded by an accurate cleaning and filtering of the point cloud [22]. Even if, thanks to the point cloud, it is possible to obtain the articulated primordial structure of the historical structures;, this lacks additional specifications on the attributes, important for the control operations. In order to integrate additional data on geometry, material and attributes, it is therefore necessary to transform the point clouds into parametric 3D geometric models [21, 23].

HBIM is a working environment where all this is possible. The difficulty in management, however, lies in the exchange of information between the parts that make up the project. The Industry Foundation Class (IFC) helps to fill this gap and therefore improve the dialogue between the parties and the swap of information.

Thanks to the IFC, data on work are given hierarchically and geometric and non-geometric data can be integrated, managed and shared among the different participants.

Parametric 3D modelling of historical buildings allows to store information about cultural, social, environmental, maintenance and other artefacts [24–26]. In addition, intangible heritage characteristics can be integrated into the 3D model in a more coherent and structured way. All data sources are combined into one database that allows easy access and extraction of information. A parametric historical BIM model containing semantic information can serve as a database. Information can be continuously expanded, edited, exported and updated during conservation, repair, and maintenance (CRM) activities.

3 Research Methodology and Data

3.1 Research Aim

The research was focused on the investigation of the current state in the field of management and maintenance of historic buildings in Slovakia and the degree of knowledge, respectively, the use of Heritage Building Information Modelling (HBIM) tools within the facility management of historic buildings. The goal was to analyse approaches, methodologies to the management and maintenance of buildings in Slovakia. The research analyses general information related to the management and maintenance of historic buildings and specifically focuses on the level of knowledge of the term Heritage Building Information Modelling.

3.2 Data Collection and Research Sample

Organizations and companies operating in Slovakia were approached through an online questionnaire survey. The aim of the survey was to address a wide professional public, i.e. companies in the public and private sector that own or manage historic buildings. As part of the survey, approximately 80 organizations were contacted by email. Of the total number of approached companies and organizations, the return of the questionnaire was approximately at the level of 22.89%. The reason for the relatively low return was mainly the reluctance, lower level of information and high workload of the respondents.

3.3 Research Step and Methodology

At the beginning of the questionnaire, the respondent was informed about the purpose of the questionnaire. The priority goal was the analysis of the current situation in the field of facility management (administration and maintenance of buildings) for historical buildings. The questionnaire consisted of 21 questions and was divided into three parts:

- Section 1: information about the respondents – type/type of organization (ownership), sector of operation (region), type of historical building
- Section 2: information on building management and maintenance – the method of maintenance executor, organizational integration of maintenance executors within the structure of the organization, the structure and format of building management and maintenance documents
- Section 3: implementation of HBIM tools in practice – degree of knowledge of the concept of HBIM, use and degree of implementation of HBIM tools, definition of obstacles and advantages of applying HBIM principles

4 Results: The Current Situation in the Field of Management and Maintenance of Historical Buildings

4.1 Information About the Respondents

The research was carried out in October 2022. The first step in the research was to obtain basic information about the respondents. From this point of view, the following data were obtained:

- The region in which the historical building is located
- The owner of the historic building
- Status/character of the historical building – cultural monument, year of declaration, protected monument in the UNESCO list
- A type of historical building

In terms of the region in which historical buildings are located, the following regions were most represented: Prešov Region (41.7%), Košice Region (33.3%) and Žilina Region (16.7%). From the point of view of the owner, the most represented were historical buildings owned by the church (66.7%), municipalities (16.7%) and historical buildings owned by the state, respectively by private owners (natural persons) - both at the level of approx. 8.3%. 75% of historical buildings have been declared cultural monuments, and 8.3% of buildings have been included in the UNESCO (United Nations Educational, Scientific and Cultural Organization) list. From the point of view of the type of historical building, churches (50%) and monasteries (16.7%) were the most represented. This was followed by town houses, wooden houses, castles or chateaux and mansions (all at the level of approx. 8.3%).

4.2 Facility Management of Historical Buildings

The second part of the research was the analysis of facility management (administration and maintenance) of buildings. Respondents answered the following questions:

- Who manages the facility management of the historic building (internal component of the administrator, external organization – external contractor)?
- If facility management is under your direction, you have created an independent department, a separate job position – administrator/building administration; facility manager?
- How many people take care of the management and maintenance of a historic building?
- Have you ever come across the term facility management?
- Do you have a building protection and maintenance plan – building use manual?
- As part of the manual for the use of the building, have you drawn up rules on use?
- As part of the manual for the use of the building, have you elaborated rules on technical inspections?
- As part of the building use manual, do you have rules for building maintenance and repairs?
- As part of the manual for the use of the building, is the timetable for planned maintenance processed?

As many as 90.9% of the respondents manage the administration and maintenance of the historic building under their own control, that is, they do not use the services of external suppliers. It was interesting to find that only 18.2% of respondents who manage the administration and maintenance of the building within their organization (internally) have created a separate department, a separate job position. On average, approximately 0–4 people (83.3%) and 5–10 people (16.7%) take care of building management and maintenance. A surprising finding was in the area of knowledge of the term facility management, where none of the respondents had met the mentioned term and did not know its definition.

Another part of Sect. 2 was focused on the elaboration of the manual for the use of the building and its individual parts. As many as 66.7% of respondents stated that they have developed a summary manual for the use of the building (summary of documents – rules for the use of the building, rules on technical inspections, rules for building maintenance and repairs, timetable for planned maintenance). The individual administrators had prepared mostly only some of the mentioned documents, specifically: in the next part of Sect. 2, we focused on the development of the manual for the use of the building and its individual parts. As many as 66.7% of respondents stated that they have developed a summary manual for the use of the building (summary of documents – rules for the use of the building, rules on technical inspections, rules for building maintenance and repairs, timetable for planned maintenance). Individual administrators had mainly prepared only some of the above documents, namely:

- Only **18.2%** of the surveyed respondents had developed **rules for the use of the building** (the rules for the use of the building consist of two separate parts – construction and technical-technological part). The construction part contains precise instructions for loading, handling, cleaning and organization of movements in the space. In part technical-technological devices, the principles of how to operate these devices in a trouble-free, safe and economical manner are precisely defined. These principles are precisely defined by operating regulations, which contain instructions for use.
- **36.4%** of the surveyed respondents had developed **rules on technical inspections** (in this section, technical inspections are precisely planned). It is defined when the given inspection will be carried out, and it is defined which parameters will be monitored on individual construction parts and technical and technological equipment. The priority of these inspections is to determine the current condition and determine the degree of damage, which serves as a prevention against the occurrence of possible malfunctions).
- **36.4%** of the surveyed respondents had developed **rules on technical inspections** (in this section, technical inspections are precisely planned). It is defined when the given inspection will be carried out, and it is defined which parameters will be monitored on individual construction parts and technical and technological equipment. The priority of these inspections is to determine the current condition and determine the degree of damage, which serves as a prevention against the occurrence of possible malfunctions).
- Only **18.2%** of the interviewed respondents had developed **up building maintenance and repair rules** (the protection and maintenance plan contains a complete technological repair/maintenance procedure (regular-preventive maintenance, induced so-called forced maintenance – emergency situation), with breakdown by individual work tasks, evaluation the need for the number of workers for the performance of individual work tasks and the evaluation of the need for material for the performance of individual work tasks).
- And approximately **27.3%** of respondents had prepared a **time schedule of planned maintenance**. Detailed breakdown of when individual planned maintenance tasks are to be performed (season, month, intensity, nature of inspection, etc.).

4.3 Degree of Knowledge of the Concept Heritage Building Information Modelling (HBIM) and Facility Management of Historical Building

After obtaining information about the respondents and the facility management of the historical building, the next part of the research was focused on the level of knowledge of the term Heritage Building Information Modelling, respectively, Facility Management of historical building. None of the interviewed respondents knew the concept of HBIM (Heritage BIM), and on the basis of this, even the interviewed respondents did not use it in the management and maintenance of their historic building. After familiarizing the respondents with the mentioned concept and the possible advantages and barriers of using HBIM in the management and maintenance of residential buildings, we found out which of the mentioned advantages and barriers the respondents consider to be decisive, significant features/facts. The respondents cited the following answers as the greatest advantages:

- Faster problem solving (80% of respondents)
- Maximizing time and money savings (70% of respondents)
- Improvement of building life cycle management (50% of respondents)
- Improvement of preventive maintenance planning (50% of respondents)
- The possibility to continuously expand, modify, export and update information during protection, repair and maintenance activities (40% of respondents)
- Improvement of client satisfaction (30% of respondents)

Among the biggest barriers to the implementation of HBIM in their processes, the respondents mentioned above all:

- High level of data complexity (77.8% of respondents)
- Lack of qualified HBIM – FM employees (66.7% of respondents)
- Solving initial data quality problems (33.3% of respondents)
- Lack of standards and guidelines (33.3% of respondents)
- The need for training (33.3% of respondents)
- Software and hardware costs (33.3% of respondents)
- Training costs and HBIM consultant (33.3% of respondents)

The conclusion of the research was focused on the willingness to implement HBIM tools and principles into their processes. To the above question, up to 72.7% of the respondents said that they are considering using the above tools (HBIM) in their activities.

5 Conclusion

The application of innovative tools in the administration and maintenance of historic buildings represent a revolutionary way of performing work within the framework of facility management. In Slovakia, traditional methods of building management and maintenance are used so far for facility management. As a priority, individual administrators emphasize the development of rules for technical inspections and the establishment of their time schedule. By implementing HBIM (Heritage Building Information Modelling) tools, it is possible to optimize individual processes, improve preventive maintenance planning, reduce costs, streamline and speed up problem solving, and many other benefits that help increase the quality of individual work and processes. The survey showed that the level of knowledge of the concept of HBIM among managers of historic buildings in the territory of the Slovak Republic is at a low level. The low level of knowledge of the term HBIM also stems from the relatively low level of knowledge of the term BIM itself. In Slovak construction practice, the rate of BIM realization in individual processes is very low. The main reasons are reluctance to change the usual ways of implementing individual processes and actions, low level of information, reluctance to learn new things, etc.

Despite the mentioned facts, the survey showed a willingness to implement uniform tools and concepts in individual activities (72.7% of respondents), but at the same time pointed out possible disadvantages and barriers preventing the rapid implementation of the mentioned tools and procedures in the processes of preparation, planning and management of facility management of buildings.

The goal of future research will be to carry out similar, extended research in the field of administration and maintenance of historical buildings in the territory of the Slovak Republic and abroad. The idea will be to establish cooperation with foreign organizations (monument offices, universities, sleeping organizations) and to carry out research in the field of knowledge and implementation of HBIM within the processes of administration and maintenance of historic buildings. Future surveys will be conducted in the form of online questionnaires and guided personal interviews.

Acknowledgements The paper presents partial research results of project APVV-17-0549 "Research of knowledge-based and virtual technologies for intelligent designing and realization of building projects with emphasis on economic efficiency and sustainability". This article was created thanks to support under the Operational Program Integrated Infrastructure for the project National infrastructure for supporting technology transfer in Slovakia II – NITT SK II, co-financed by the European Regional Development Fund.

References

1. Barazzetti, L., Banfi, F., Brumana, R., Gusmeroli, G., Previtali, M., & Schiantarelli, G. (2015). Cloud-to-BIM-to-FEM: Structural simulation with accurate historic BIM from laser scans. *Simulation Modelling Practice and Theory, 57*, 71–87. Retrieved from:https://www.sciencedirect.com/science/article/abs/pii/S1569190X15000994. Accessed 2022/10/15.
2. Fai, S., Graham, K., Duckworth, T., Wood, N., & Attar, R. (2011, September 12–16). Building information modelling and heritage documentation. In *Proceedings of the 23rd international symposium, international scientific committee for documentation of cultural heritage (CIPA), Prague, Czech Republic*, pp. 12–16. Retrieved from https://www.researchgate.net/publication/308634697_Building_Information_Modeling_and_Heritage_Documentation. Accessed 2022/10/15.
3. Ashworth, S., & Tucker, M. (2017, August). *FM awareness of building information modelling (BIM)*. Retrieved from https://www.researchgate.net/publication/319159345. Accessed 2022/10/15.
4. Gallaher, M. P., O'Connor, A. C., Dettbarn, J. L., Jr., & Gilday, L. T. (2004). *Cost analysis of inadequate interoperability in the U.S. capital facilities industry* (p. NIST GCR 04-867). National Institute of Standards and Technology. Retrieved from https://nvlpubs.nist.gov/nistpubs/gcr/2004/nist.gcr.04-867.pdf. Accessed 2022/10/16.

5. Eastman, C., Teicholz, P., Sacks, R., & Liston, K. (2008). *BIM handbook: A guide to building information modeling for owners, managers, designers, engineers, and contractors*. John Wiley & Sons. ISBN 978-0-470-18528-5. Retrieved from https://www.wiley.com/en-sg/BIM+Handbook:+A+Guide+to+Building+Information+Modeling+for+Owners,+Designers,+Engineers,+Contractors,+and+Facility+Managers,+3rd+Edition-p-9781119287537. Accessed 2022/10/16.
6. IFMA. *What is facility management?* Retrieved from https://www.ifma.org/about/what-is-fm/. Accessed 2022/10/16.
7. UNITED-BIM.com. How BIM is facilitating facility management process? Retrieved from https://www.united-bim.com/how-bim-is-facilitating-facility-management-process/. Accessed 2022/10/17.
8. Becker, F. (1990). *The total workplace*. Van Nostrand.
9. Moore, M., & Finch, E. (2004). Facilities management in South East Asia. *Facilities, 22*, 259–270.
10. IFMA. *11 core competencies of facility management*. Retrieved from https://ifmacdn.azureedge.net/sfcdn/docs/default-source/marketing/pd-pages/11-core-competencies_oct2020.pdf?sfvrsn=2. Accessed 2022/10/17.
11. UNITED-BIM.com. *How BIM is facilitating facility management process?* Retrieved from https://www.united-bim.com/how-bim-is-facilitating-facility-management-process/. Accessed 2022/10/18.
12. eSUB- Construction Software. (2018). *How to improve operational efficiency in construction with BIM modelling*. Retrieved from https://esub.com/blog/how-to-improve-operational-efficiency-in-construction-with-bim-modeling/. Accessed 2022/10/18.
13. Murphy, M., McGovern, E., & Pavia, S. (2009). Historic building information modelling (HBIM). *Structural Survey, 27*, 311–327. Retrieved from https://www.researchgate.net/publication/241582141_Historic_building_information_modelling_HBIM. Accessed 2022/10/18.
14. Heesom, D., Boden, P., Hatfield, A., De Los Santos Melo, A., & Czarska-Chukwurah, F. (2020). Implementing a HBIM approach to manage the translocation of heritage buildings. *Engineering, Construction and Architectural Management, 28*, 2948–2966. Retrieved from https://wlv.openrepository.com/handle/2436/623683. Accessed 2022/10/18.
15. Yang, X., Grussenmeyer, P., Koehl, M., Macher, H., Murtiyoso, A., & Landes, T. (2020). Review of built heritage modelling: Integration of HBIM and other information techniques. *Journal of Cultural Heritage, 46*, 350–360. Retrieved from https://www.researchgate.net/publication/342929270_Review_of_built_heritage_modelling_Integration_of_HBIM_and_other_information_techniques. Accessed 2022/10/18.
16. Godinho, M., Machete, R., Ponte, M., Falcão, A. P., Gonçalves, A. B., & Bento, R. (2019). BIM as a resource in heritage management: An application for the National Palace of Sintra, Portugal. *Journal of Cultural Heritage, 43*, 153–162. Retrieved from https://www.sciencedirect.com/science/article/abs/pii/S1296207419305710. Accessed 2022/10/18.
17. López, F. J., Lerones, P. M., Llamas, J., Gómez-García-Bermejo, J., & Zalama, E. (2018). A review of heritage building information modeling (H-BIM). *Multimodal Technologies and Interaction, 2*, 21. Retrieved from https://www.mdpi.com/2414-4088/2/2/21. Accessed 2022/10/19.
18. Sharafat, A., Khan, M. S., Latif, K., Tanoli, W. A., Park, W., & Seo, J. (2021). BIM-GIS-based integrated framework for underground utility management system for earthwork operations. *Applied Sciences, 11*, 5721. Retrieved from https://www.mdpi.com/2076-3417/11/12/5721. Accessed 2022/10/19.
19. Mill, T., Alt, A., & Liias, R. (2013). Combined 3D building surveying techniques-Terrestrial laser scanning (TLS) and total station surveying for BIM data management purposes. *Journal of Civil Engineering and Management, 19*, 23–32. Retrieved from https://www.researchgate.net/publication/261643973_Combined_3D_building_surveying_techniques-Terrestrial_laser_

scanning_TLS_and_total_station_surveying_for_BIM_data_management_purposes. Accessed 2022/10/19ň.

20. Bruno, S., De Fino, M., & Fatiguso, F. (2018). Historic building information modelling: Performance assessment for diagnosis-aided information modelling and management. *Automation in Construction, 86*, 256–276. Retrieved from https://www.sciencedirect.com/science/article/abs/pii/S0926580517301164. Accessed 2022/10/19.
21. Sanhudo, L., Ramos, N. M. M., Martins, J. P., Almeida, R. M. S. F., Barreira, E., Simões, M. L., & Cardoso, V. (2020). A framework for in-situ geometric data acquisition using laser scanning for BIM modelling. *Journal of Building Engineering, 28*, 101073. Retrieved from https://www.sciencedirect.com/science/article/abs/pii/S2352710219314470, Accessed 2022/10/20.
22. Jalón, M. L., Chiachío, J., Gil-Martín, L. M., & Hernández-Montes, E. (2020). Probabilistic identification of surface recession patterns in heritage buildings based on digital photogrammetry. *Journal of Building Engineering, 34*, 101922. Retrieved from https://www.sciencedirect.com/science/article/abs/pii/S2352710220335543. Accessed 2022/10/20.
23. Dinis, F. M., Sanhudo, L., Martins, J. P., & Ramos, N. M. M. (2020). Improving project communication in the architecture, engineering and construction industry: Coupling virtual reality and laser scanning. *Journal of Building Engineering, 30*, 101287.
24. Quattrini, R., Pierdicca, R., & Morbidoni, C. (2017). Knowledge-based data enrichment for HBIM: Exploring high-quality models using the semantic-web. *Journal of Cultural Heritage, 28*, 129–139. Retrieved from https://www.sciencedirect.com/science/article/abs/pii/S1296207417300754. Accessed 2022/10/20.
25. Quattrini, R., & Baleani, E. (2015). Theoretical background and historical analysis for 3D reconstruction model. Villa Thiene at Cicogna. *Journal of Cultural Heritage, 16*, 119–125. Retrieved from https://www.sciencedirect.com/science/article/abs/pii/S1296207414000223. Accessed 2022/10/20.
26. Nadkarni, R. R., & Puthuvayi, B. (2020). A comprehensive literature review of multi-criteria decision making methods in heritage buildings. *Journal of Building Engineering, 32*, 101814. Retrieved from https://www.sciencedirect.com/science/article/abs/pii/S2352710220334471. Accessed 2022/10/21.

Index

L. Knapcikova, S. Zohrehvandi (eds.), *9th International Conference on Mobility, IoT and Smart Cities*, EAI/Springer Innovations in Communication and Computing,
https://doi.org/10.1007/978-3-031-28225-6